# FIRE ALARM SYSTEM

***By Cornel Barbu &Juliana Barbu***

**ISBN:** 978-1-4583-1619-6

**ID#9967478**

**www.lulu.com**

**Dedicated to my wife, Julia**

# CONTENT

## Why I am writing this book

When I first began my career in electrical construction, I was surprised to see that many electricians, technicians and even engineers found it difficult when it came time to design, estimate or install the fire alarm system (FAS) for the projects they were designing, quoting or building. Why was this the case? The answer was easy :

Fire alarm system (FAS) is relatively new and quickly need to be up-graded as changes are required due to advances in technology. As a result these systems remain complicated for electricians as well as designers and estimators. The point is you need to think ahead and be a self-starter. The fact that you bought this book is a sign that you're a self-starter. People such as you don't let anything hold you back: you learn everything you can and perform very well in any field. They experience real success and are well-suited for almost any field.

This book is written to provide you with authentic, useful information about fire alarm systems, about the materials and the steps to follow during installation, and to provide all the necessary details to understand how a fire alarm system operates. It also covers what are the manadatory or minimal conditions that need to be considered whan it comes to quoting, installing and/or maintaining such a system.

In order to investigate the subject of fire alarm systems we need to have a ***project*** in our discussion. Via the project which we will address each step item by item until we reach the main goal which is:

*To provide an individual (whether they have an electrican's licence or not) with enough information to understand and build a fire alarm system for any residential, commercial or industrial project.*

## The scope of fire alarm systems (FAS)

When a fire occurs in any type of building a system must be in place that alerts and protects the equipment, people and property within that facility.When a fire alarm signal is activated people will leave the building. The system then provides information to the authorized personnel about the event. The system also prevents or slows smoke and fire spreading through the auxiliary devices.A fire is a major event. The scope of the Fire Alarm System is to anticipate and prevent damage and injury being incurred from the fire and to maintain safety when such an event occurs. Important information is provided about the system through the FACP (fire alarm control panel).Several systems operate and are interconnected with the fire alarm system. These are:

- The sprinkler system
- Elevators
- Emergency power supply (generator)
- HVAC devices
- Access control system

An interlocking logic diagram usually needs to be in place for this. It is a special requirement and, as an electrician, you should know this. I will mention a few of the interlocking functions:

1. A fire alarm system should monitor and control the water pressure of the sprinkler system. It's a disaster if there is a fire and not enough water distributed from the sprinkler system. This means the water pressure must be properly controlled and monitored. Devices for the fire alarm and sprinkler systems therefore need to be integrated during installation.
2. In case of fire in a high-rise or mid-rise building, the FAS should generate a signal to bring the elevator to the first floor (homing) or to where it is required by the fire department.
3. The vital loads-into facility of the generator's monitoring system should be interconnected with the FAS.
4. If there is a fire inside any building with an HVAC system, the make-up air units (supply) are controlled by the FAS. Smoke inside ducts for any reason is a fire event and will be controlled by the FAS.
5. During a fire, door holders are released so no area remains closed and no one is trapped.

## Project identification

*To provide someone interested in the electrician's trade with enough information to understand and install a fire alarm system for a residential, commercial or industrial project,* we need to create the "scope of work" first.

I will select a "design" based from my experience and try to cover all the possible scenarios you'll need to be aware of. You may already be working with these systems at your workplace if you are estimating electrical projects or doing electrical construction and/or maintenance work.

Our working model will consist of a two-level building with a basement designated for institutional/commercial use. Utilities within the building include: electricity, gas, water, heating, ventilation and air conditioning, etc.
As far as spaces to house technology, there will be: an electrical room, sprinkler room, an elevator room and a diesel generator (provides emergency power supply).
In terms of working spaces, the building consists of offices, meeting rooms and classrooms/conference rooms. The building itself is non-combustible which means the structures, frames and beams are metal.Inside the building, the partitions are created on metallic frames and drywall.

In our design model we assume there is a legal mandate to request have a working Fire Alarm System in the building. Therefore we will focus on fully explaining this system. In order to ensure that you have a thorough understanding of the system, we will cover "touch" tasks like calculations , explanations of where and when to perform certain tasks related to the FAS and include a large range of diagrams and illustrations. This will help your understanding and retention of the material.
However, if you have any questions at all as you read through this book, please send me an e-mail at: axaelectric@yahoo.ca or go to: www.lulu.com and search at book sections the key word : electrician's book or
This text will mainly reference and discuss material from the point of view of the electrical trade and will not cover aspects of civil or mechanical engineering.
A diagram of the building will be included as part of our "hands-on" training and learning process regarding the Fire Alarm System (FAS).
When necessary details and notes will be included to clarify tasks and concepts and hopefully eliminate confusion.
Let's now see what the building in which we need to install our FAS looks like!

1. Ground floor arrangements

There are special rooms at this level, such as:

- Electrical Room & Sprinkler Room
- Elevator Room & Storage room

Arrows indicate the access area into the building. The corridor area is highlighted and refers to rooms R115 and R101. A stairway (R116) is located in room R101. The building will have a controlled access system to provide access into the building at the Main Entrance (R 100). The east, north and southwest doors are emergency doors and are locked to prevent entry only. The Main Entrance door can be locked and access to the interior of the building is controlled by: a card reader (CR), electric strike (ES) or a magnetic lock (MH). These integrate with the Fire

Alarm System. When the FAS is activated, it signals the release of the magnetic holders which keeps the doors unlocked.

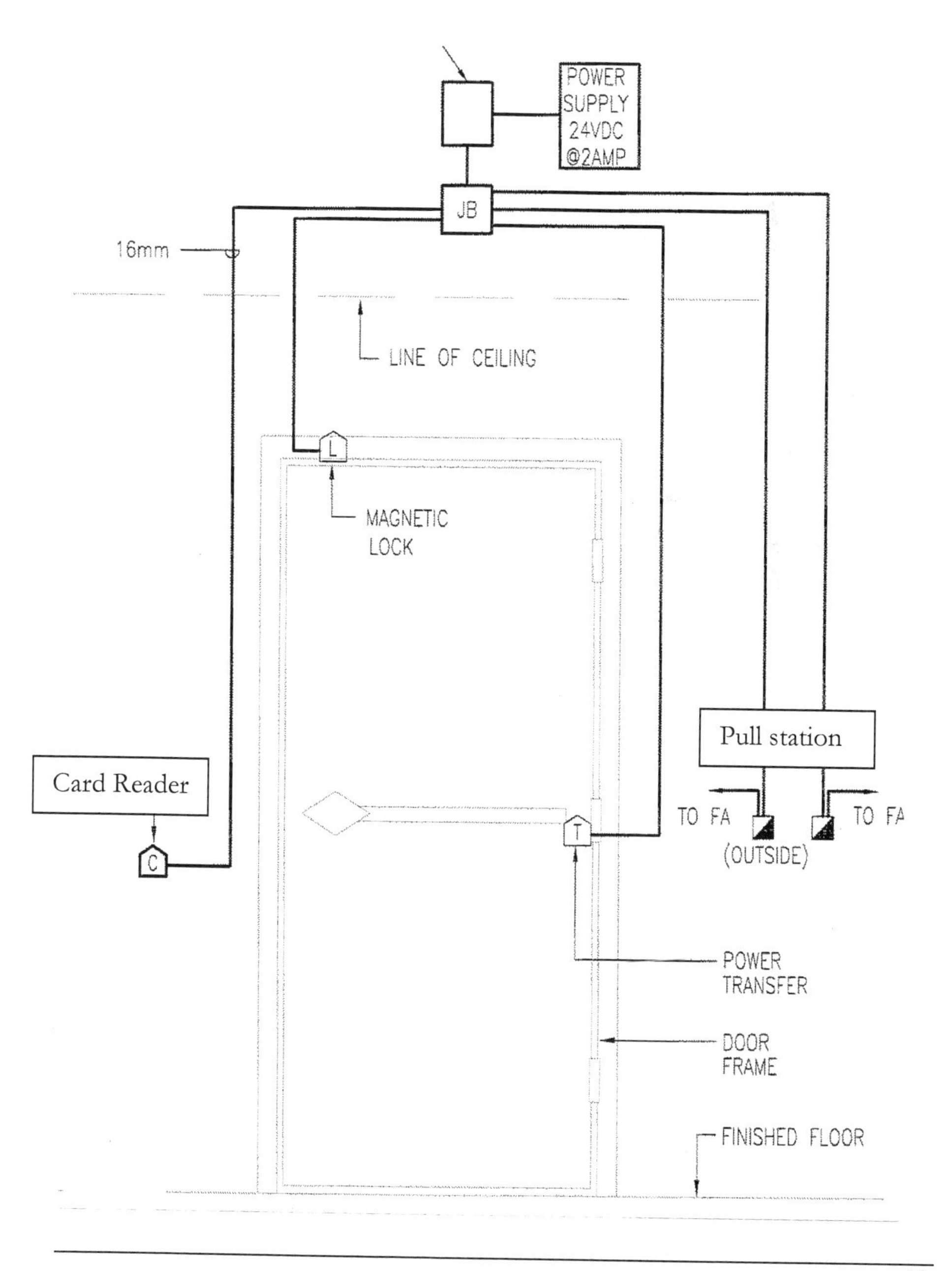

Now let go back to project descriptions:

Ground Floor room schedule:

| Item # | Room # | Description |
|---|---|---|
| 1 | R100 | Main Entrance |
| 2 | R101 | South West Corridor |
| 3 | **R103** | **Electrical Room** |
| 4 | **R105** | **Sprinkler Room** |
| 5 | R107 | Office |
| 6 | R109 | Men's Washroom |
| 7 | R111 | Women's Washroom |
| 8 | R113 | Office |
| 9 | R102 | Office |
| 10 | R104 | Class Room 1 |
| 11 | R106 | Class Room 2 |
| 12 | **R108** | **Elevator Room** |
| 13 | R110 | Class Room 3 |
| 14 | R112 | Class Room 4 |
| 15 | **R114** | **Storage Room** |

The designation of the rooms is important for our study model as they relate to the fire alarm system. For example, in the **Sprinkler Room (R105)** there is a typical requirement for the installation of a fire alarm system. Here the fire alarm system should have devices from both circuits:

1. INI –initiation circuit and
2. NAC –notification and annunciation circuit.

The NAC (devices like horn/strobes/speakers) will notify occupants of the need to evacuate the area should a emergency/fire event occur on the system. The INI (devices like detectors, pull-station or monitoring & control modules) will initiate signals for the FACP (fire alarm control panel) that are integrated with the sprinkler system:

- Low/High water pressure on the system
- Valves supervision( valves status : open /close)
- Status of the Flow meters

<u>This gives information about the availability and integrity of the system</u>. It is supposed keep the system "under control" so it can "react" when needed. A non-responsive or "faulty" system or parts of the system is a major problem. Therefore, maintenance is crucial. The next few pages will explain how systems such as the sprinkler system are interconnected with the FAS.

2. First floor arrangements

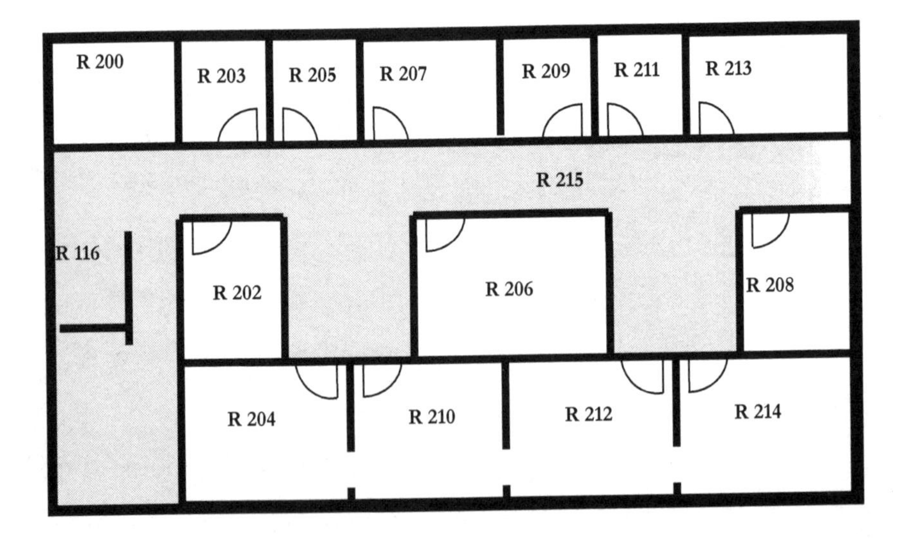


The designation or locations of the rooms is important for our study model since they are related to the fire alarm system. For example, in the **Elevator Room (R206) and Stairway (R 216)** we have the typical requirements for the installation of a fire alarm system. Here the fire alarm system should have devices from both circuits: the INI –initiation circuit and NAC – notification and annunciation circuit.

The NAC (devices like horn/strobes/speakers) will notify us about any event in an area so people will know to evacuate the area.

The INI (devices like detectors, pull-station or monitoring & control modules) will initiate signals for the FACP (fire alarm control panel) in relation to the elevator/elevator machine room and stairway. During a fire or emergency, people will be instructed to use stairways during the evacuation, as there are restrictions with regard to the use of elevators. In most buildings when there is an alarm from the FACP, the elevators are automatically directed to go a specific location (level). This is known as the "homing" status. The elevator shaft and elevators should be supervised by the FAS via the proper devices. The stairwells should also be supervised by the FAS as they are the only way to evacuate the building during a fire or emergency. Others disciplines like civil engineering will focus on such areas as they are required to include in their designs ways ensure these maintain structural integrity and are safe during disasters like fire and earthquakes. This gives information about the availability and integrity of the system. It is supposed keep the system "under control" so it can "react" when needed. A non-responsive or "faulty" system is a major problem. Therefore, maintenance is crucial. The next few pages will demonstrate the integration of the FAS with the elevator shafts and stairwells.

First Floor room schedule:

| Item # | Room # | Description |
|---|---|---|
| 1 | R200 | Office |
| 2 | R201 | South West Corridor |
| 3 | R203 | Office |
| 4 | R205 | Office |
| 5 | R207 | Office |
| 6 | R209 | Men's Washroom |
| 7 | R211 | Women's Washroom |
| 8 | R213 | Office |
| 9 | R202 | Office |
| 10 | R204 | Class Room 1 |
| 11 | R206 | Elevator Room |
| 12 | R208 | Storage Room |
| 13 | R210 | Class Room 2 |
| 14 | R112 | Class Room 3 |
| 15 | R114 | Class Room 4 |

## 3 Basement floor arrangements

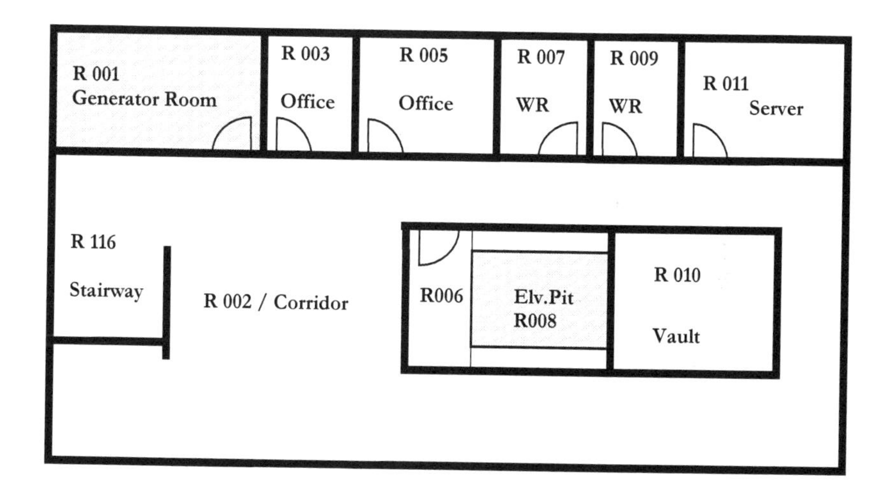


## Basement floor schedule

| Item # | Room # | Description |
|---|---|---|
| 1 | **R001** | **Generator Room** |
| 2 | R003 | Office |
| 3 | R005 | Office |
| 4 | R007 | Men's Washroom |
| 5 | R009 | Women's Washroom |
| 6 | R011 | Server Room |
| 7 | R110 | The Vault |
| 8 | R006 | Elv.Machine Room |
| 9 | R008 | Elevator Pit |
| 10 | R002 | Corridor |
| 11 | R116 | Stairway |

The room designation is important for our study model since it is related to the fire alarm system. For example, in the Generator Room (R001) and Elevator Pit and Machine Room (R006), there are typical requirements for the installation of a fire alarm system.

Here the fire alarm system should have devices from both the INI –initiation circuit and NAC –notification and annunciation circuit.

The NAC (devices like horn/strobes/speakers) will notify occupants about any event in an area in order to begin evacuation. The INI (devices like detectors, pull-stations or monitoring & control modules) will initiate signals for the FACP (fire alarm control panel) relating to the elevator/elevator machine room and stairway. During a fire or other emergency, people are directed to use the stairs during evacuation as there are restrictions regarding the use of elevators and the generator should be ready to supply power to the vital loads. For example, imagine we are talking about a hospital. There are systems in which the power is shut-off during a fire but in other cases the generator should be able to automatically start and supply the vital loads.

In most buildings the generator is monitored and automatically starts.

The status of the generator is indicated also by the FAS

- generator running or
- Low fuel level or
- The generator's circuit is in "trouble"

The information mentioned above needs to be monitored in order to provide dates about the availability and integrity of the generator in these areas where is required.

It is supposed keep the system "under control" so it can "react" when needed. A non-responsive or "faulty" system is a major problem. Therefore, maintenance is crucial. The next few pages will demonstrate how the Generator Room is interconnected with the FAS.

## Conventional Fire Alarm Systems

The conventional Fire Alarm System the type of system usually found in small projects and meets minimum safety requirement at an acceptable price. This was the way Fire Alarm Systems were initially designed. However, over time, such systems have been upgraded due to advances in technology and design considerations.The fire alarm system's main component is the control panel. The control panel is the place where the alarm and supervision signal originates. As soon as a fire or false alarm is initiated by these devices  via the smoke/heat detectors or pull-stations in the building, the control panel will initiate an acoustic or visual notification into the area. These devices are connected to the fire alarm control panel (FACP). Such a FACP normally has two types of circuits:

1. **Initiation Circuits (INI )**
2. **Notification and Annunciation  circuit(NAC)**

The integrity of the circuits (INI and NAC) must be maintained in order to prevent defect s, such as: wiring being cut-off; isolation of the wiring or damage; wires disconnected or accidental grounding of the circuits. Such supervision of the panel ensures that the system will be reliable and work properly when needed. To account for this, designers provide a solution for the system so the integrity of the circuits is consistently maintained.

This solution is known as the EOL resistor.

EOL means End of Line.

If, when the system is in operation, a defect appears on the INI or NAC circuit, this supervisory function will detect this as "trouble" and will give details about the nature and type of defect. This is important as people expect such systems to properly work when necessary in order to protect lives and property.

This diagram displays this:

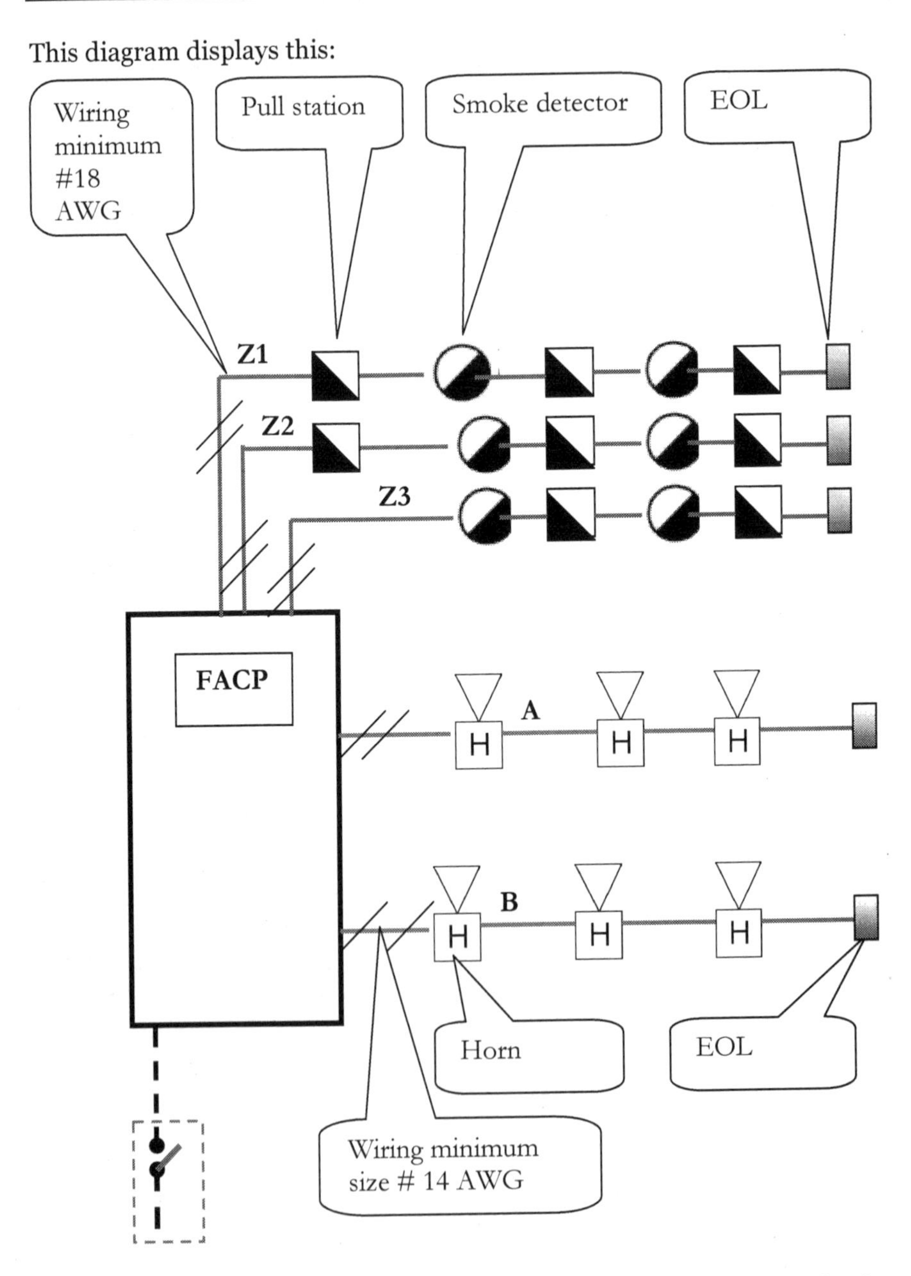


This is the Designated Circuit breaker for the power supply (120 volts). This circuit breaker is located in the power supply electrical panel, which will be locked. It is a requirement that most electrical codes be painted in RED.

The Initiation Circuit (INI) includes devices that will trigger the system to work. The parts of the initiation circuit are: detectors, pull stations and end of line resistors. Since the initiation circuits are limited by the number of devices and are designed to cover different or specific locations, they are scheduled by zones Z1; Z2; Z3 in our picture.

The Notification and Annunciation circuit (NAC) includes devices that send out notifications about a fire alarm event (horns; strobes; combination horn/strobe; speakers). The notification and annunciation circuits are limited by the number of devices. For safety reasons, it is necessary to include in this same controlled area multiple devices from different circuits. This increases the reliability of the system. In case where one circuit that is supposed to notify occupants about a fire or other emergency event fails for whatever reason, the other circuit will still be available. C1; C2 are shown below in the illustration.

This is the way the EOL is supervising the circuit's integrity :

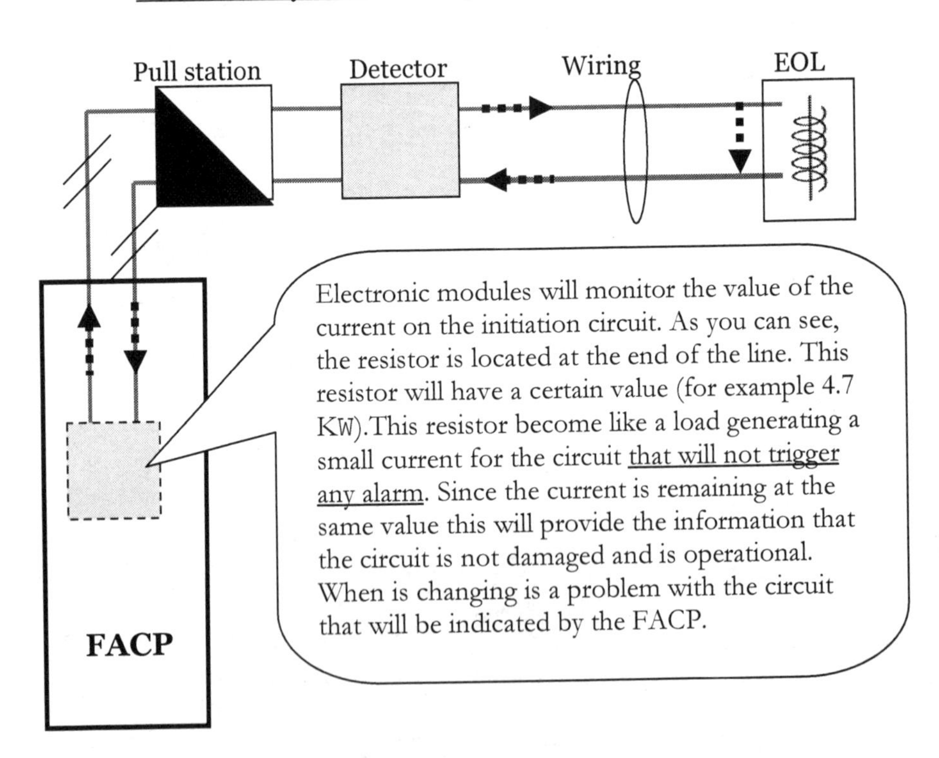

## Addressable Fire Alarm System

The Addressable Fire Alarm System is the system usually found in big projects. The addressable fire alarm system appeared as an upgrade of conventional fire alarm system. It includes devices like: detectors, pull stations and an individual address that is unique and easy recognized by the fire alarm control panel (FACP). The fire alarm system's main component is the control panel. The FACP (fire alarm control panel) is where the alarm and supervision signal originates. As soon as a defect or emergency is detected devices like detectors or pull-stations, an acoustic or visual notification will be initiated into the area. These devices are connected to the fire alarm control panel (FACP) using wiring.

Such a FACP normally has two types of circuits:

1. **Initiation Circuits (INI )**
2. **Notification and Annunciation Circuit(NAC)**

The integrity of the circuits (INI and NAC) must be maintained in order to prevent defect s, such as: wiring being cut-off; isolation of the wiring or damage; wires disconnected or accidental grounding of the circuits. Such supervision of the panel ensures that the system will be reliable and work properly when needed. To account for this, designers provide a solution for the system so the integrity of the circuits is consistently maintained. This solution is known as the EOL resistor. EOL means End of Line.

If, when the system is in operation, a defect appears on the INI or NAC circuit, this supervisory function will detect this as "trouble" and will give details about the nature and type of defect. This is important as people expect such systems to properly work when necessary in order to protect lives and property.

This diagram displays them:

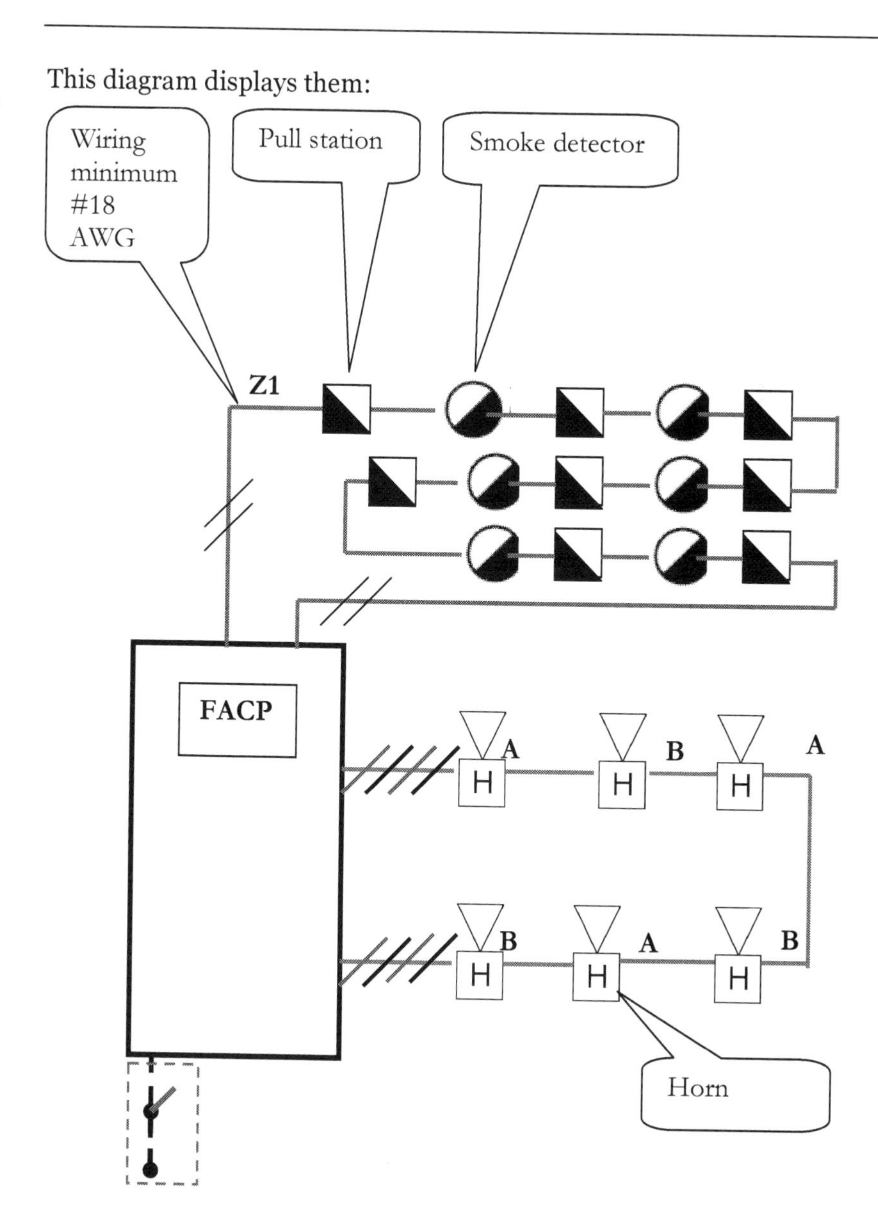


This is the Designated Circuit breaker for the power supply (120 volts).This circuit breaker is located in the power supply electrical panel; is lockable and most electrical codes require that it be painted in RED.

**The Initiation Circuit (INI)** includes devices that trigger the system to work. The parts of the initiation circuit are: detectors, pull station and end of line resistors.

Since the initiation circuits are limited by the number of devices and designed to cover different or specific locations they are scheduled by zones. Only one zone is shown in Z1.

**The Notification and Annunciation circuit (NAC)** includes devices that send out notifications about a fire alarm event. These devices are: horns; strobes; combination horn/strobe and speakers. The notification and annunciation circuits are limited by the number of devices. For safety reasons, it is necessary to include in this same controlled area multiple devices from different circuits. This increases the reliability of the system. In case where one circuit (A or B) that is supposed to notify occupants about a fire or other emergency event fails for whatever reason, the other circuit will still be available. A&B are shown in our diagram.

The EOL is located in the FACP. It will have the same designation as indicated in the Conventional Fire Alarm System.

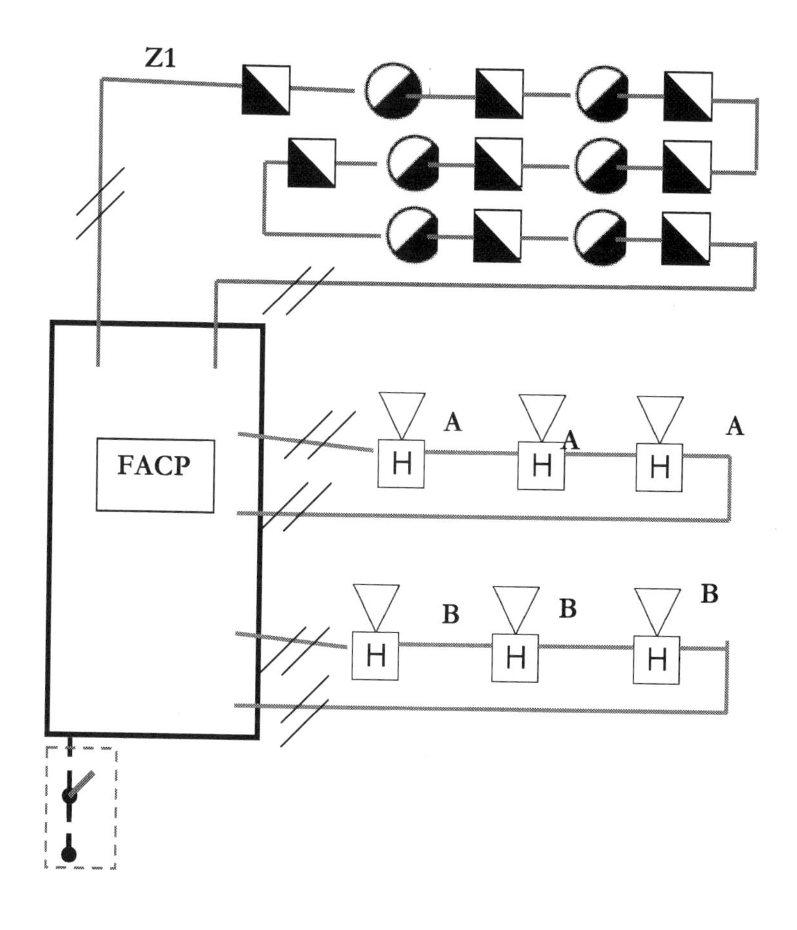


The Designated Circuit breaker is for the power supply (120 volts). This circuit breaker is located in the power supply electrical panel; it is lockable and most electrical codes require that it be painted in RED.

In such a system, devices like detectors and/or pull stations are identified and are able to provide information regarding the system's integrity, alarm status, fault status, etc. This means needing such devices to communicate with the main FACP, which is like the brain of the system. Because it has a unique address, the addressable device will provide information about the location the emergency or event is happening and is connected by wiring in a loop. This usually consists of two wire loops. The system will have more than one loop since the area to be supervised or controlled is generally quite large. The number of loops is a function of the maximum number of devices that can be placed on one loop. Initiation circuits that use low current devices (electronic ones). For example, with a GE (old Edwards) device you can place at least 250 devices on one loop. This includes "intelligent" devices like Isolating modules, Input/Output modules or Relay modules. These will be explained further in upcoming chapters.

This system is used globally, is very easy to install and requires less maintenance. These are major advantages. In the USA and Canada, manufacturers like Mircom and GE Security (Edwards) introduced this system and I will recommend any of their suppliers since this system is complete and provides good, reliable products for a Fire Alarm System.

Addressable devices are "intelligent" devices as they are able to provide alarm notification in advance of a fire happening. Such devices detect when smoke is present or when the heat level for specific equipment rises above the prescribed level.

The practice is that the supplier of the system will do the system programming at the commissioning stage. Most of the time the end user (the client) will be advised to have these maintained and checked by a certified verification company periodically after the system is operational.

Considering the liability involved with fire alarm systems, most countries have requirements in place that "regulate" this field. Some of these include:

- Who is doing the installation
- Who is doing the verifications
- Who and how the end user is maintaining the system

An electrician may be involved in any of these abovementioned steps so is important that you know this.

In conclusion, an addressable system for an FAS contains devices that own and address and communicate with the central panel. Having an address programmed into the system will inform occupants as to WHERE (location or ZONE) the event is happening and the type of the event. Imagine a high-rise building with 40 levels. In the event of a fire on the 36th floor you'll be able to know such specifics as the very room in which the event is occurring.

This is important when it comes to evacuation and the control and extinguishing of the fire by the fire department. It ensures greater safety of property and life.

The addressable loop wiring class A is a two-wire loop with both ends coming back to the FACP. These interconnect "intelligent" initiation devices which contain an individual address to provide an individual response.
In this way every device (detector, pull station, module, or duct detector) will have a designated address that corresponds to a designated location.
This address has a specific function that will generate a specific task.
Having the FACP attached to software makes it possible to program a system that is able to supervise, initiate and notify in an efficient and sophisticated manner during any fire event.

## Fire Alarm Control Panel (FACP)

As you can see the sketch bellow, Fire Alarm Control Panel (FACP) is interconnected with other systems, such as:

Telephone system,
Security system,
Electrical system,
Sprinkler system and
Various mechanical equipment (HVAC).

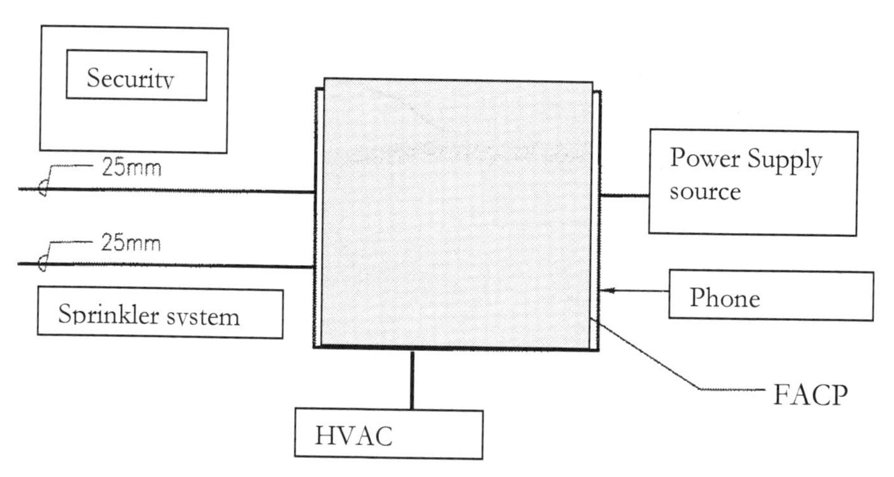


Most of the time, the FACP is located at the entrance to a building. The reason for this is that the FACP needs to be accessible as it displays crucial information concerning a fire emergency and the integrity of the fire alarm system. There are situations in which it is not possible to install the FACP at the entrance of a building. In these cases, **annunciation panels** will be installed at the entrance to make the information from the FACP available. The role of the FACP is to monitor the status of the devices located in field, the integrity of the circuit, and to provide an error-free display of all events, devices and circuit statuses.

The FACP will indicate:

- System normal
- Ground fault
- Trouble
- Open circuit
- Zones status

In some instances, the location in which the FACP's box is installed should be away from the sprinkler or be a sprinkler proof type of box

## Annunciation panel

If there are situations where is not possible to install the FACP at the entrance of the building then **Annunciation panels** will be installed at the entrance to make the FACP information available. This panel is providing information for firefighters and maintenance persons about the status of the system, the status of the circuits and the location of an emergency or fire event. These panels usually need two wires for power (12 or 24 volt +& -) and 2 pairs of twisted wire for data, 16AWG in size. These needs to be run in separate conduits. Read the manufacturer's instructions to save time and to get answers of any questions you may have. The annunciation panel should be easily visible, accessible and, in the case where there are two entrances to a building, another annunciation panel should be located at the second entrance. Firefighters need to be able to gather this information from the system before entering the building. It is better to have a remote panel that is made by the same manufacturer that provided the fire alarm control panel to avoid problems with the integration into the system. For example, if the FACP is a product of GE Secutity (Edwards) and the annunciation panel is a product of Mircom or Simplex there could be problems regarding the compatibility between the systems. Situations like this could be find when you're required to make an extension for an existing system or an up-grade. The Annunciation Panel will provide information and indicate the ON/OFF status of the:

- Power
- Fire Alarm
- Ground Fault
- Trouble

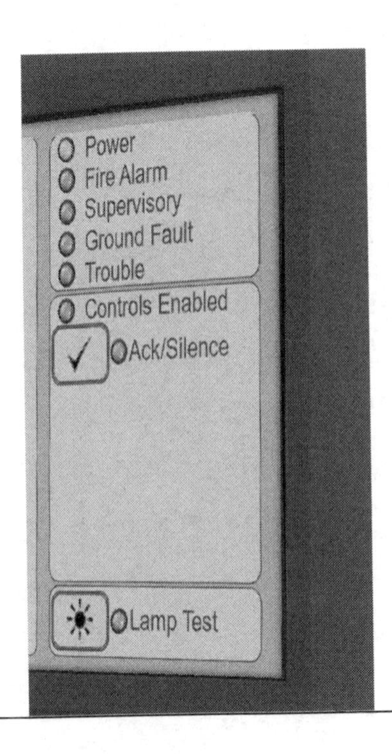

Look at the portion of the annunciation panel that displays the status of the system!

Look at the portion of the annunciation panel that displays the messages and status of the system!

## Small sketch to understand the FACP and Annunciation Panel location

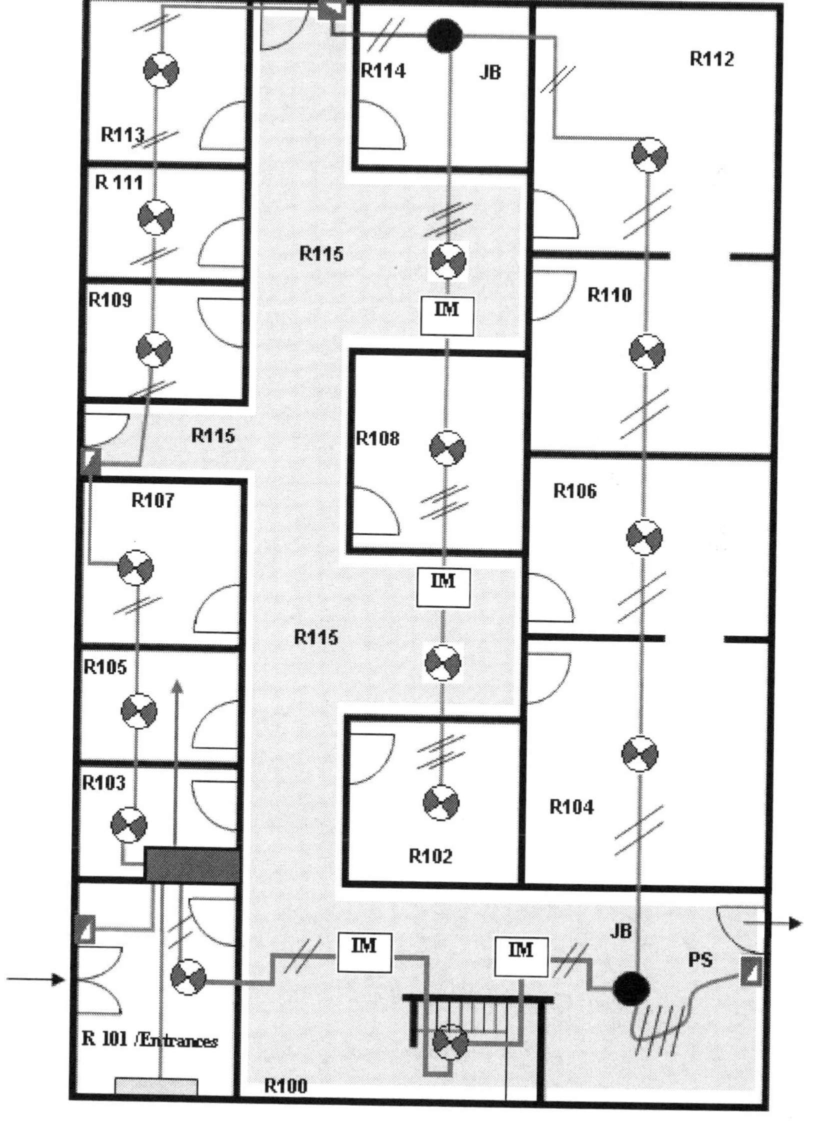


**FACP** in room R103 (right side of the door into electrical room)

**Ann. Panel** in room R101 (right side of the entrance door)

## Fire Alarm System: Components

The main components of the FAS system are:

1. Fire Alarm Control Panel (FACP – this will include the software and the alarm and notification modules)
2. Annunciation panel (an extension of the FACP's display & control)
3. Initiating Circuits (INI - that will generate the alarm)
4. Notification and Annunciation circuits (NAC- will warn /notify people with acoustic and visual signals about an emergency)
5. End –of –Line resistor (EOL – able to monitor the integrity of INI & NAC circuits)

There are components that are auxiliary to the system and these are not supervised by the FACP but they are interconnected with FACP. Such auxiliary devices are:

- Door holder releases (Magnets on the wall to hold door)
- Fan shut down devices
- Door magnetic lock release
- Elevator (recall function)
- Pressurization fans

## Single Stage & Two Stage Fire Alarm Systems

Single Stage Fire Alarm Systems:

This is the system where smoke, overheating, fire, etc. causes an alarm to be triggered so that the evacuation of the building can commence. These are the most common situations you'll need to have managed with such a system.

Two Stage Fire Alarm Systems:

In some types of buildings a false alarm could cause panic since the number of people needing to be evacuated is much larger. In the case of a false alarm, it can cause a huge mess.

Designers provided a "two stage system" that basically will be generated in the case of an event or alarm that brings the FACP into ALERT status.

This signal will be intercepted by the maintenance personnel or monitoring company. At this point the authorized personnel will start investigating the event indicated by the FACP to see if it is a fire or simply a false alarm.

If there is evidence of real fire or serious event, the authorized personnel will have a special key. By inserting the key into any pull stations (located most of the time close to the doors and provided with the key switch) the alert mode of the FACP is changed to EVACUATION status. An evacuation will commence.

In the situation where the FACP is still in ALERT status and a second alarm appears, the panel will switch automatically into EVACUATION status.

There is secret here is that the two stage FAS will have a manual pull station with a key switch to be used by authorized personnel when intervention is required. So now.................... you can make the difference between single stage FAS and two stage FAS.

These are the characteristics of the Two Stage FAS.

This is the KEY SWITCH

This is a Manual Pull Station that needs to be installed into a Two Stage FAS (observe the key switch where the arrow is).

## More about Initiation circuit

## Class B initiation circuit

The Initiation Circuit (INI) includes devices that trigger the system to work. The parts of the initiation circuit are: detectors, pull stations and end of the line resistors.

Since the initiation circuits are limited by the number of devices and designed to cover different areas (locations) they are scheduled by zones:Z1; Z2; Z3 in our picture.

The picture shows Initiation circuit (INI) **as class B circuit.** As you observe the lines will not come back to the panel - requirement is for a class B type of installation. The connection is a 2-wire cable connection.

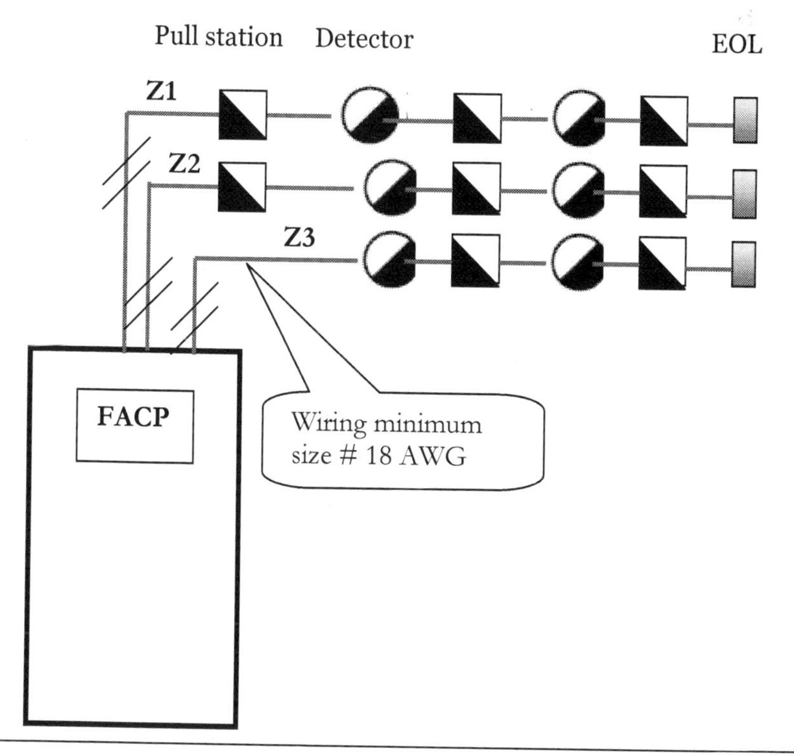

## More about Initiation circuit

## Class A initiation circuit

The sketch will follow is representing Initiation circuit (INI) as **class A circuit** and the lines will come back to the panel - the requirement is for a class A type. The connection is a 2-wire cable connection.

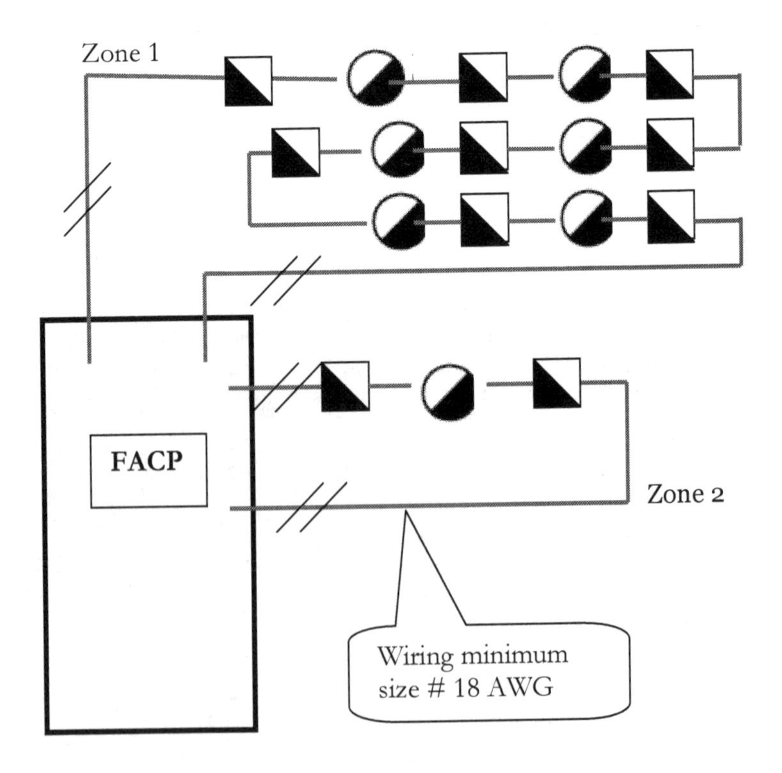


Such panels could have more than 2 zones. (Zone 1 and Zone 2) Over time addressable devices (with individual address) and such systems became more complex, flexible and easy in regards of maintenance. Technology had a real impact and the following symbols list will indicate some of these advances. Three main new items appeared:

- Addressable control module
- Addressable monitor module
- Addressable isolation module

You might find a lot of symbols to represent the devices in FAS.

| SYMBOL | DESCRIPTION |
|---|---|
| FIRE ALARM | |
| ⊘ | SMOKE DETECTOR |
| ⊘ R | SMOKE DETECTOR c/w (4 WIRE BASE) RELAY BASE |
| (FS) | FIRE ALARM SPEAKER c/w VISUAL SIGNAL |
| [FS] | FIRE ALARM SPEAKER c/w VISUAL SIGNAL |
| [FACP] | FIRE ALARM CONTROL PANEL |
| | PULL STATION |
| | ADDRESSABLE MONITOR MODULE |
| | ADDRESSABLE CONTROL MODULE |
| S D | DUCT SMOKE DETECTOR |
| FS | FLOW SWITCH |
| SV | SUPERVISED VALVE |

Try to identify the abovementioned devices in the diagram on this page. We can observe a loop form the FACP to the first device and coming back to the FACP. It is class A circuit as mentioned in the title. Later we will go into detail about such a loop. The main thing is to understand what class A circuit is.

Typical class "A"- Initiation **circuit (INI)** as a loop.

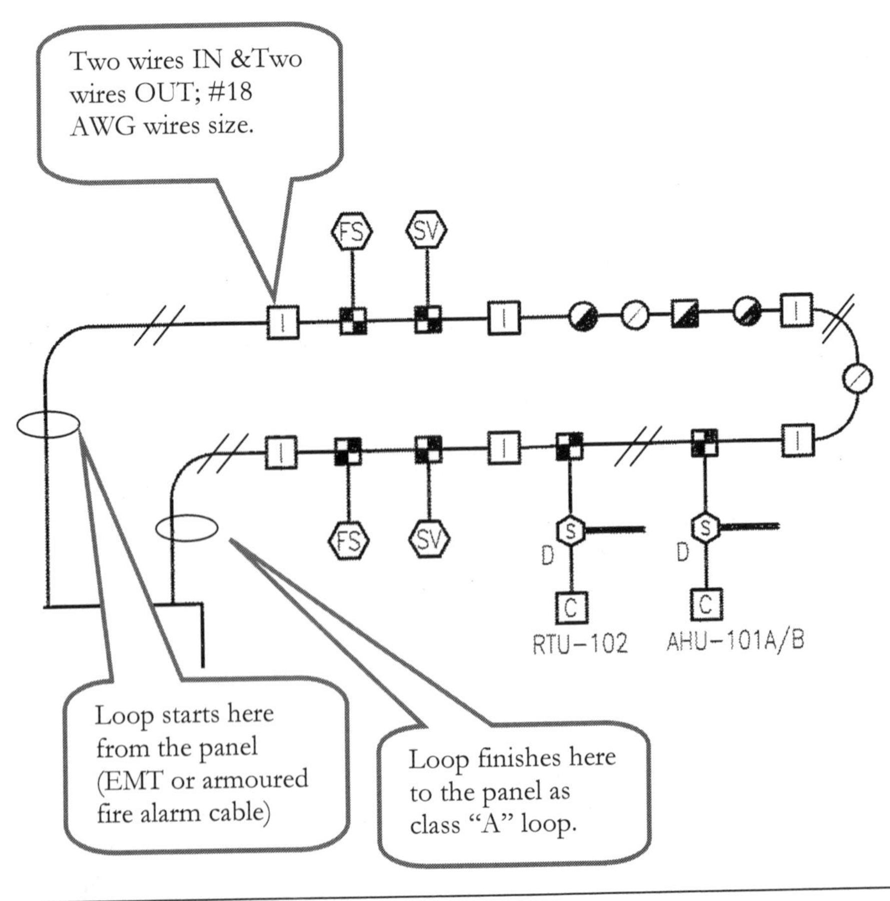

In order to make sure the initiation circuit is available all the time there is EOL resistor inside the FACP.
The integrity of the initiation circuits (INI) is monitored so the system will indicate when the circuit is defective or grounded or in working status. Some possible defects include:

- wiring being cut;
- damage to the wiring's insulation;
- wires are disconnected;
- accidental grounding of the circuit lines

These are unlikely to happen and for this reason the integrity of the connection lines is "constantly under observation." This will ensure a reliable system that works when required. The solution is an EOL resistor. It is important for you to understand the purpose of this resistor. The image will follow will explain in bubble the way this resistor is working......

Here is how EOL supervises the circuit's integrity :

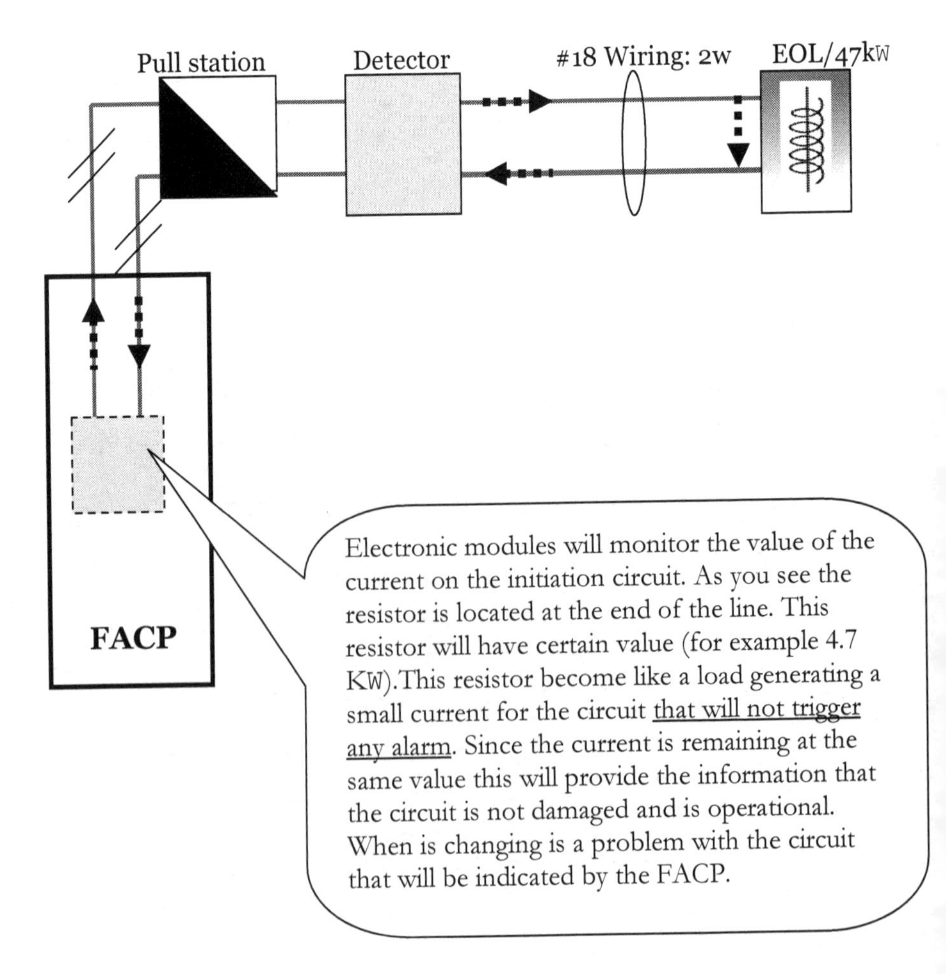

## More about Notification circuit

## Class B notification circuit

The Notification and Annunciation circuit (NAC) includes devices that send out notifications about a fire alarm event (horns; strobes; combination horn/strobe; speakers).The notification and annunciation circuits are limited by the number of devices. For safety reasons, it is necessary to include in this same controlled area multiple devices from different circuits. This increases the reliability of the system. In case where one circuit that is supposed to notify occupants about a fire or other emergency event fails for whatever reason, the other circuit will still be available.   A and B are shown in our picture.

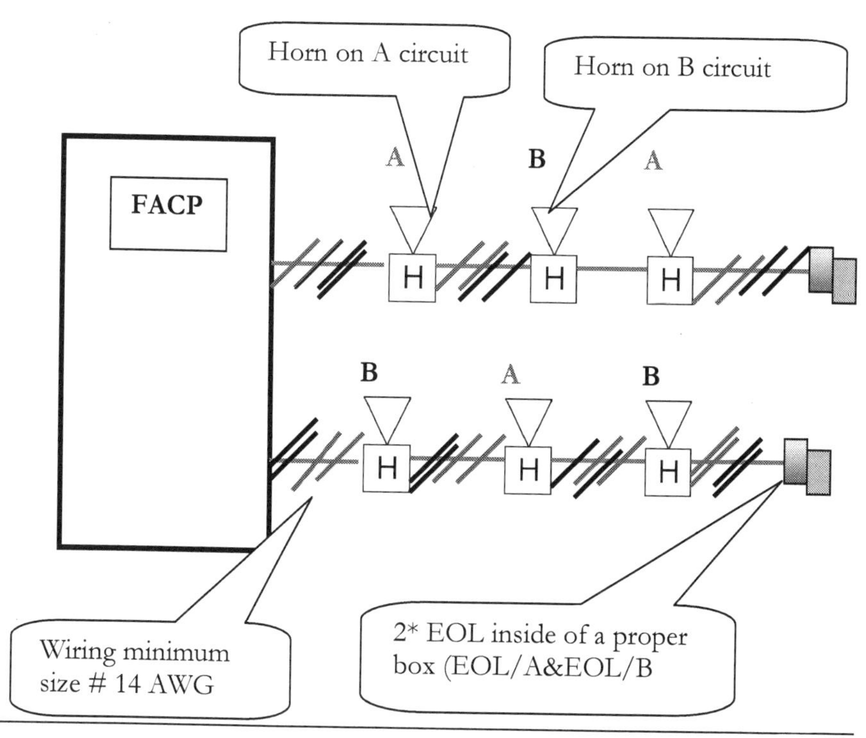

The notification circuit is a **class B circuit** and will not come back to the panel if the requirement is for a class B type.

This is 2-wire termination for the Horn/Strobe. Red is for circuit A and Black for circuit B (red wires will terminate in the H-A since black into H-B).

## More about Notification circuit

## Class A notification circuit

The notification circuit is a **class A circuit** and will come back to the panel if the requirement is for a class A type.

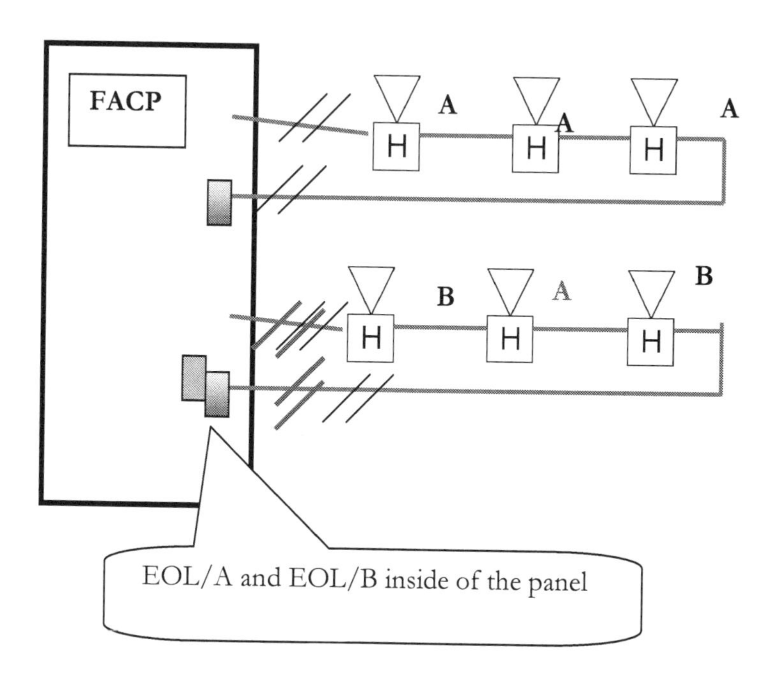


This is 2-wire termination for the Horn/Strobe. Red is for circuit A/Black for circuit B (red wires will terminate in the H-A and black into H-B).

Any adjacent (nearby, close) devices required for notification such as speakers, strobes or horns will need to be on a separate circuit. The reason for this is obvious. In a situation during which the system loses power in one of the circuits, the separate circuit will ensure that there are not large area(s) without devices to notify people of the emergency (fire, smoke, heat, etc.) as soon the alarm been triggered by the initiation circuit. Let's imagine a long corridor that has all its horns and strobes on the same notification circuit: When the alarm is triggered, if the notification circuit is compromised for any reason during the fire event that area of the corridor will have no devices working and/or will be unable to notify the public of the emergency. This would pose a major problem so please read all the drawings carefully so you'll be able to identify and avoid such a situation when doing system installations. For the notification circuit we have the same kind of circuit monitoring process as the one explained for the initiation circuit. Defects like:

- wiring cut-off;
- damage to the insulation of the wires;
- wires are disconnected;
- accidental grounding of circuit lines

These are unlikely to happen when the integrity of the connection lines is "constantly under observation." This will ensure a reliable system that works when required. The solution is an EOL resistor.

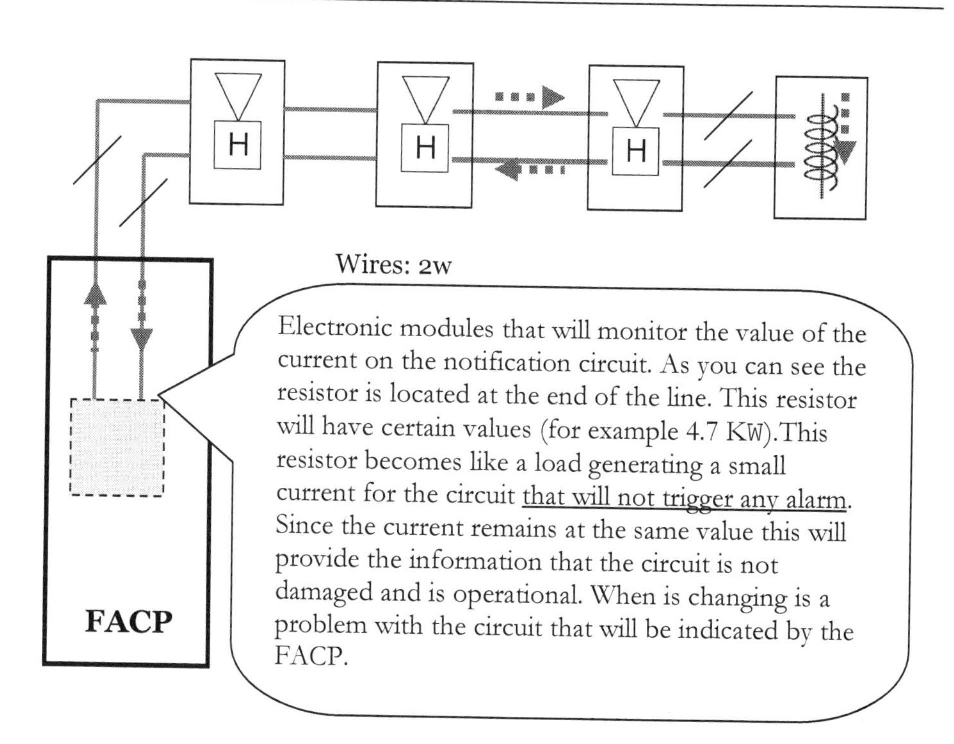
H
H
H
Wires: 2w
Electronic modules that will monitor the value of the current on the notification circuit. As you can see the resistor is located at the end of the line. This resistor will have certain values (for example 4.7 KW).This resistor becomes like a load generating a small current for the circuit that will not trigger any alarm. Since the current remains at the same value this will provide the information that the circuit is not damaged and is operational. When is changing is a problem with the circuit that will be indicated by the FACP.
FACP

## Type of cables and wiring in fire alarm system

As I mentioned before, the devices are connected with wiring or cables or combinations of wiring and cables composed of the INI and NAC circuit and then connected to the FACP.The cables could be shielded or non-shielded and the wires will be in accordance with the requirements of the designer. The wires are copper wires and their size is determined according to the current they carry (small signal currents).

For example on the INI circuits the minimum size is #18 AWG. Since the current on the NAC circuit is a bit larger, the minimum size is #14AWG.

In both kinds of circuits the numbers of devices is limited as well and there is definitely a relationship between the load and the wiring size. Distance is a factor as well because a long run between the FACP and last devices is limited due to drop voltage considerations. No drop voltage larger than 10 % is acceptable for any signal circuit. The system and devices are working with a voltage range between 24-30 volts DC. This is a consideration when you select the wiring and cables for such a system. The environment in which the system is being installed is also a factor that needs to be considered when selecting the wiring and cables.

Armored cable

Please have an example of cables specification for fire alarm system:

Specification Fire Alarm/Control Cables :

Armor Galvanized Interlocking Steel Strip (red-striped)

Conductors Solid Copper

Conductor TFN( solid) 18 & 16 AWG and/or THHN 14 & 12 AWG

Insulation

Assembly Polyester Assembly Tape; Twisted Shielded:

Laminated Aluminum/Mylar® Shield with

Tinned Copper drain wire

Maximum FPLP: 105°C (dry)

Temperature Rating MC: 90°C (dry)

Grounding one or more grounding conductors may be

bare or insulated green, see chart below

Neutral Conductor White

Maximum 300V (FPLP = fully rated plenum)

Voltage Rating 600V (MC= metal clad)

## Wiring method in fire alarm system

Wiring method refers to the way we choose to run the wiring and cables for a fire alarm system.
This system should have circuits available during adverse conditions such as: high humidity, high temperatures and mechanical stress. The system still needs to be able to inform about fire events when they occur in these conditions.
Obviously these types of conditions require the wiring be enclosed into metal conduits.
During projects I've found some PVC conduits for fire alarm systems embedded into concrete walls to ensure a high level of protection. Most of the time you will follow the designer's requirements but keep in mind that all projects require effective protection against mechanical damages, humidity and dangerous temperatures.
Please study the local building codes to find out the proper requirements for your system. When wires are protected inside conduits or different types of raceways they need to be installed in separate conduits so as not to be located in the same raceway with the wiring of other systems like the power supply, DC wiring or AC wiring (except in areas where they terminate or permitted by the designer or codes). The bottom line is to keep them in separate conduits. I would recommend keeping INI and NAC circuits in separate conduits as well.

All metal conduit or metallic raceway portions of the fire alarm system (FAS) should be bonded to grounding system. As a rule bonding wire is mandatory and needs to be installed in any conduit, box and raceway carrying fire alarm cables, wires and devices.

Be sure you understand the difference between the bonding wire and the grounding wire. This will protect people from shock hazards and grounding defects will be addressed properly. The next few chapters will help you understand this further through using our project-based learning process.

**Your notes here:**

---------------------------------------------------------------------

---------------------------------------------------------------------

---------------------------------------------------------------------

---------------------------------------------------------------------

---------------------------------------------------------------------

---------------------------------------------------------------------

---------------------------------------------------------------------

---------------------------------------------------------------------

---------------------------------------------------------------------

---------------------------------------------------------------------

---------------------------------------------------------------------

---------------------------------------------------------------------

---------------------------------------------------------------------

## The power supply for Fire Alarm System

I almost missed including this chapter but remembered it when I discussed the INI and NAC system as I made reference to the power supply of the FACP. We cannot ignore this in our discussion since the power supply issue is an important one and should be addressed into a very responsible manner.
We have a battery section in almost any FACP that acts as a reliable power supply for the system and for the notification circuit. But the main power supply, which is the AC line, will need to be designed and installed in such a way to ensure that no disconnecting or switching off of the system will happen.
You will find requirements concerning no disconnecting refers to the area between the FACP and the over current protection device for this branch circuit supplying the FACP.

**Your notes here:**

---------------------------------------------------------------------

---------------------------------------------------------------------

---------------------------------------------------------------------

---------------------------------------------------------------------

---------------------------------------------------------------------

---------------------------------------------------------------------

---------------------------------------------------------------------

---------------------------------------------------------------------

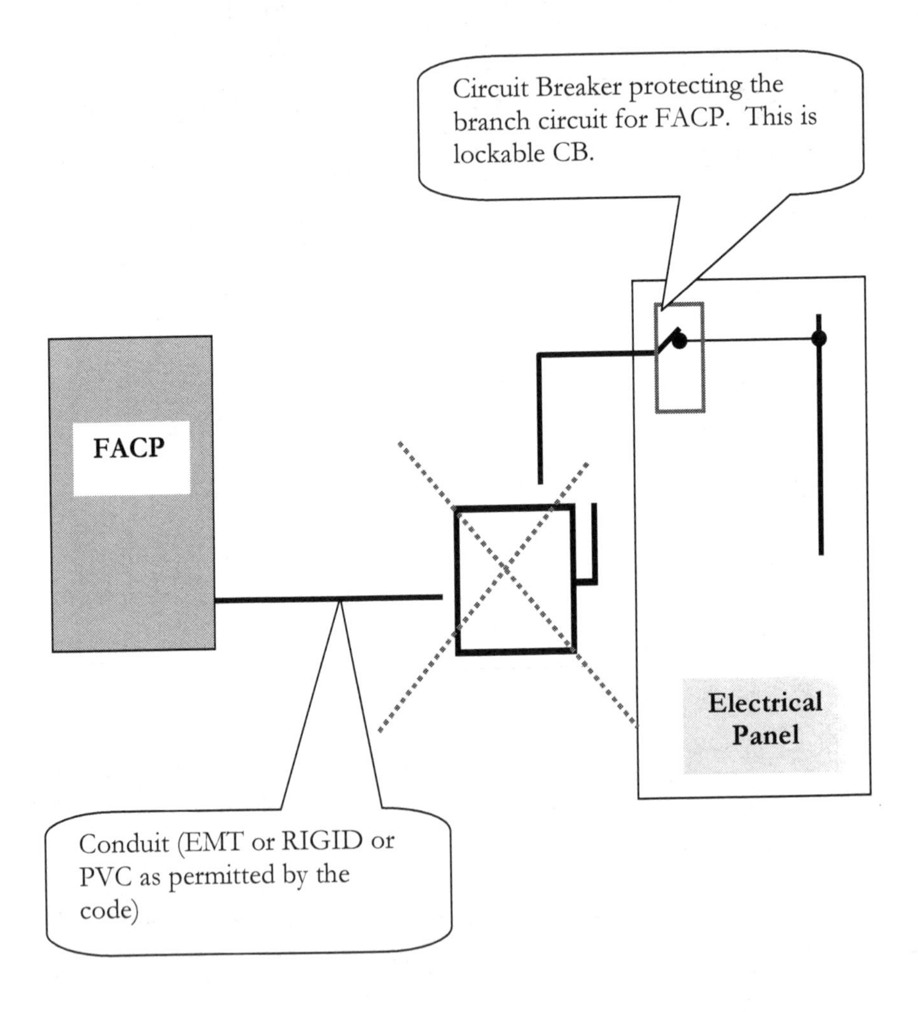
Circuit Breaker protecting the branch circuit for FACP. This is lockable CB.
FACP
Electrical Panel
Conduit (EMT or RIGID or PVC as permitted by the code)

This detail below is the right type of installation for the power supply.

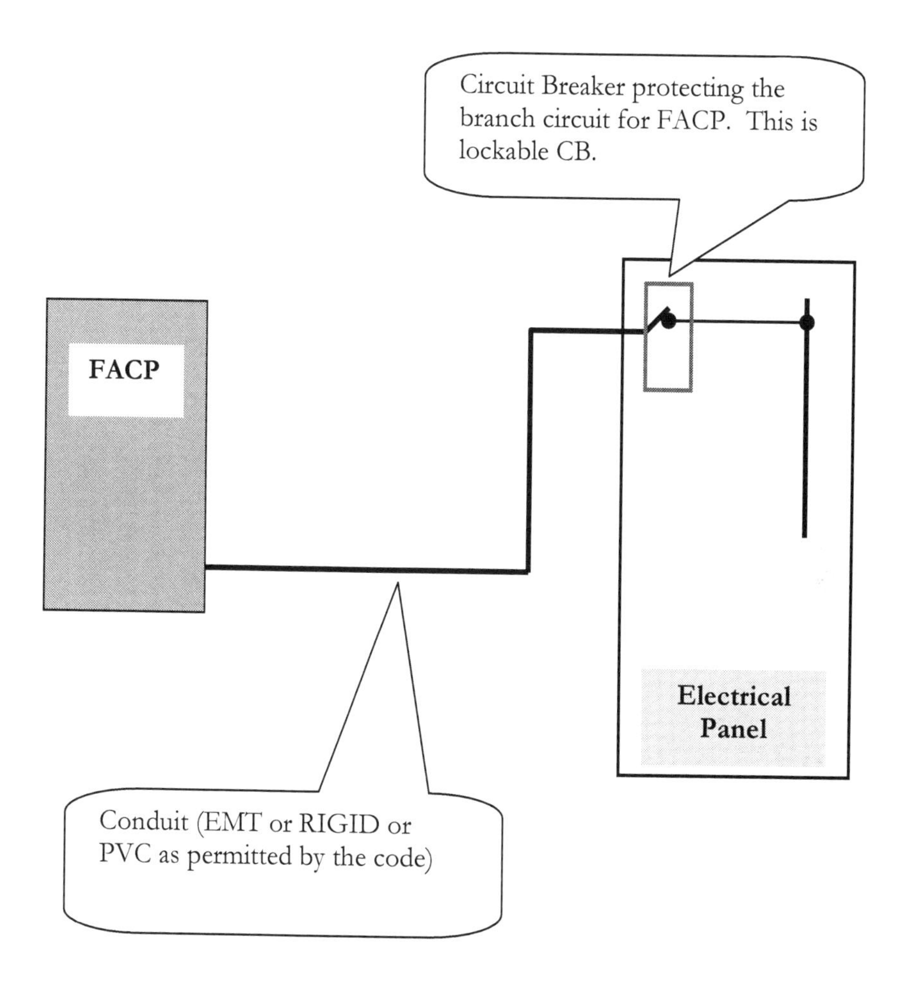


This should be a **designated circuit** for this purpose only. No other devices should be connected to this branch circuit

The over current device shall be painted red.

120/208 is the most common voltage level (USA & Canada)

## Special devices for addressable FAS

| SYMBOL | DESCRIPTION |
|---|---|
| FIRE ALARM | |
| ⊘ | SMOKE DETECTOR |
| ⊘ R | SMOKE DETECTOR c/w (4 WIRE BASE) RELAY BASE |
| FS (circle) | FIRE ALARM SPEAKER c/w VISUAL SIGNAL |
| FS (box) | FIRE ALARM SPEAKER c/w VISUAL SIGNAL |
| FACP | FIRE ALARM CONTROL PANEL |
| | PULL STATION |
| | ADDRESSABLE MONITOR MODULE |
| | ADDRESSABLE CONTROL MODULE |
| S D | DUCT SMOKE DETECTOR |
| FS (hexagon) | FLOW SWITCH |
| SV | SUPERVISED VALVE |
| I | Isolator Module |

The following page shows a typical initiation loop and the devices arranged on the loop. This is a type "A" loop that brings the signal from the field to the fire alarm control panel (FACP).

Based on the type of signal, an alarm or message will be triggered via the relay alarm or through the NOTIFICATION and ANNUNCIATION circuits. That way the scope of the system is achieved.

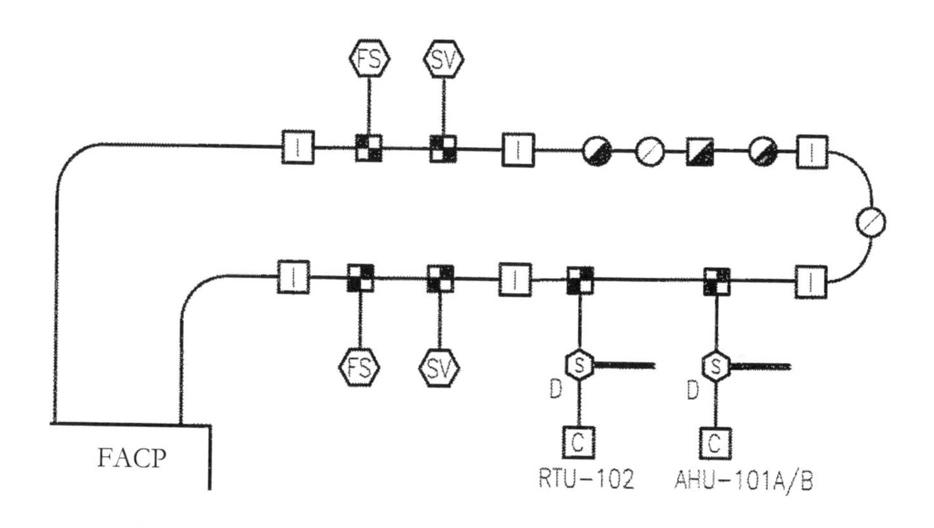


**Note:** There is no specific location for devices like detectors, manual pull stations, or monitor/control modules on the loop. This is a representation to show the interconnection between the devices into the loop. This diagram is a representation of the main types of devices found on a typical INITIATION loop.

All the special devices indicated on the diagram below represent a class "A" typical initiation (INI ) with a 2-wire loop.

- As you can see at the symbol table , at position one (1) there is a description like: SMOKE DETECTOR

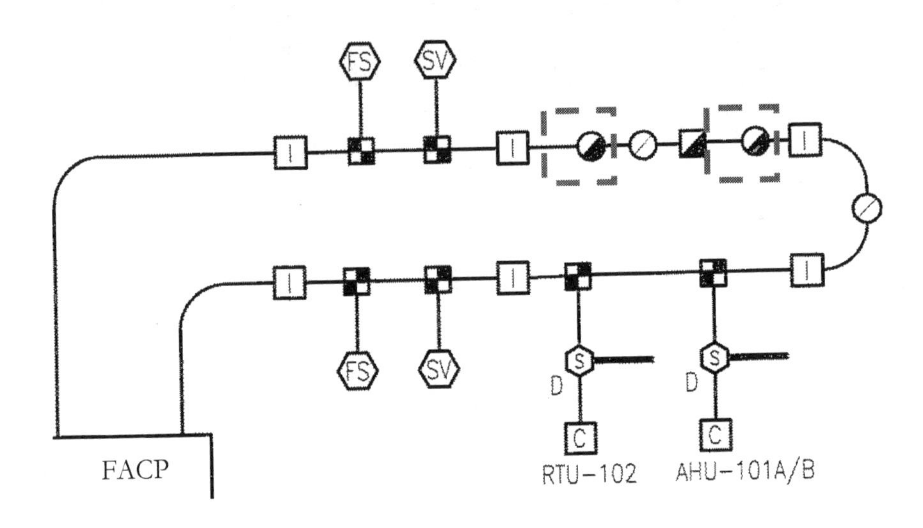


You should understand that this loop does not include only two smoke detectors. Observe the way I marked into a square box the smoke detectors as devices.
Your project will have more than two smoke detectors (the maximum number is about 250) and they are indicated in specific drawings. In the drawings you'll find them represented like they are in this diagram. Designers will usually provide you with <u>a concept diagram.</u> You'll build or estimate the system following this concept considering the entire number of detectors needed or the network of these devices.
If this information is not available in the design than you should be ready to create one taking into consideration all technical

specifications and/or specific codes requirements. In the next few pages I've included a model of a project that is intended to teach you these requirements so if you're in the situation where you are either building, designing or estimating you'll be familiar with the concept of the fire alarm system (FAS).

How the Smoke detector looks like!

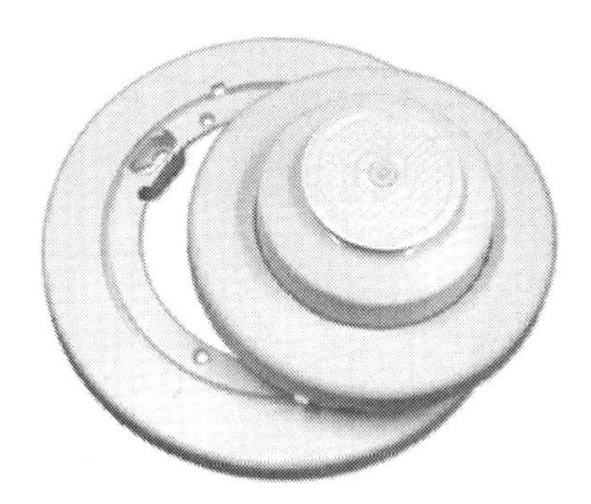

Heat detector c/w (complete with) detector base.

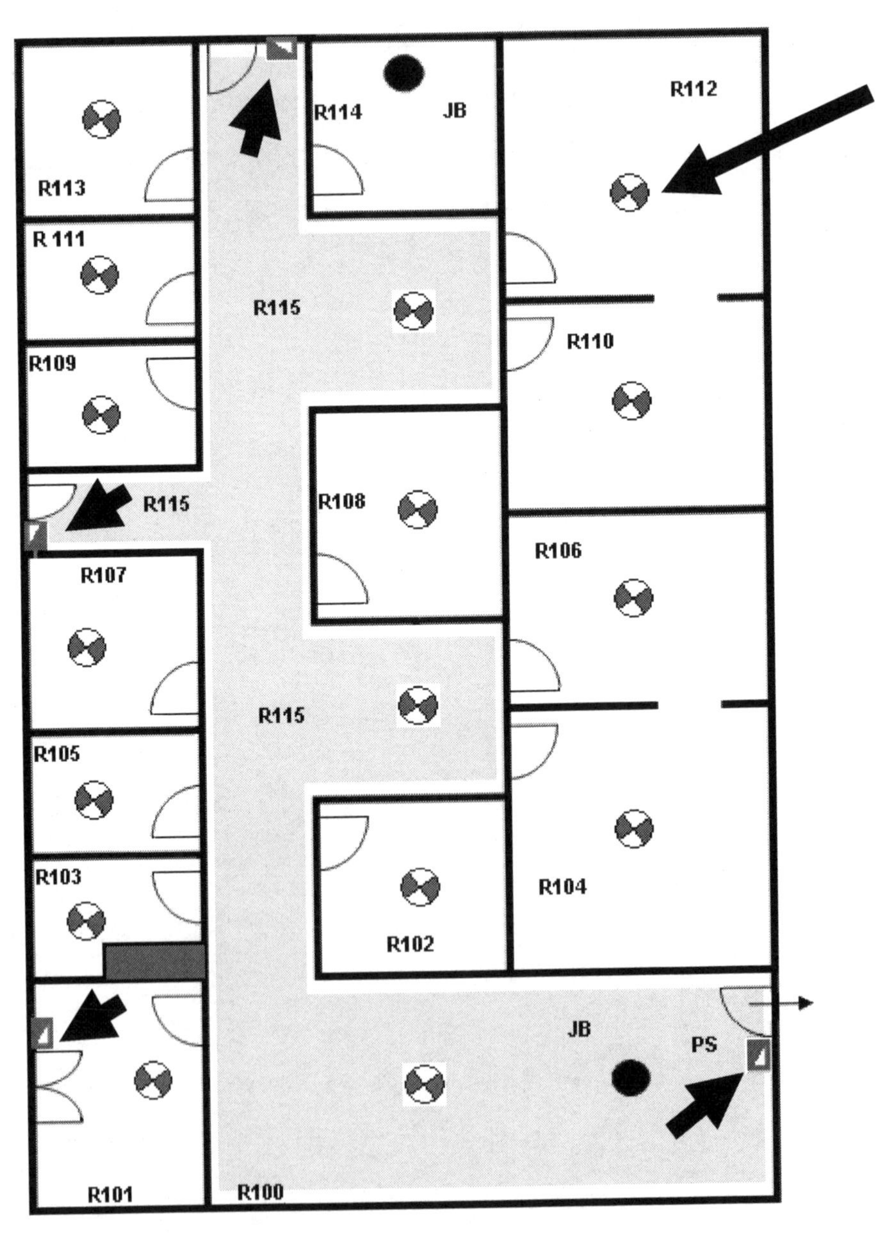


This is our model project at the main floor (grounding floor). The areas (room or corridors) where there are smoke detectors (circle with red/white) are shown.

---

The arrow (the big one) indicates a smoke detector.
On this drawing (main elevation or ground floor) we can identify about 16 smoke detectors. Pull station are located close to the escape areas (the doors). Small arrow is indicating these devices.....The black circle located in room R114 is a JB or junction box....Later we will see the type of connections between these devices and the way they complete the initiation loop.

**Your notes here:**

------------------------------------------------------------

------------------------------------------------------------

------------------------------------------------------------

------------------------------------------------------------

------------------------------------------------------------

------------------------------------------------------------

------------------------------------------------------------

------------------------------------------------------------

------------------------------------------------------------

------------------------------------------------------------

------------------------------------------------------------

------------------------------------------------------------

------------------------------------------------------------

------------------------------------------------------------

------------------------------------------------------------

------------------------------------------------------------

- Note that at position six (6) of the symbol table there is a description like: MANUAL PULL STATION

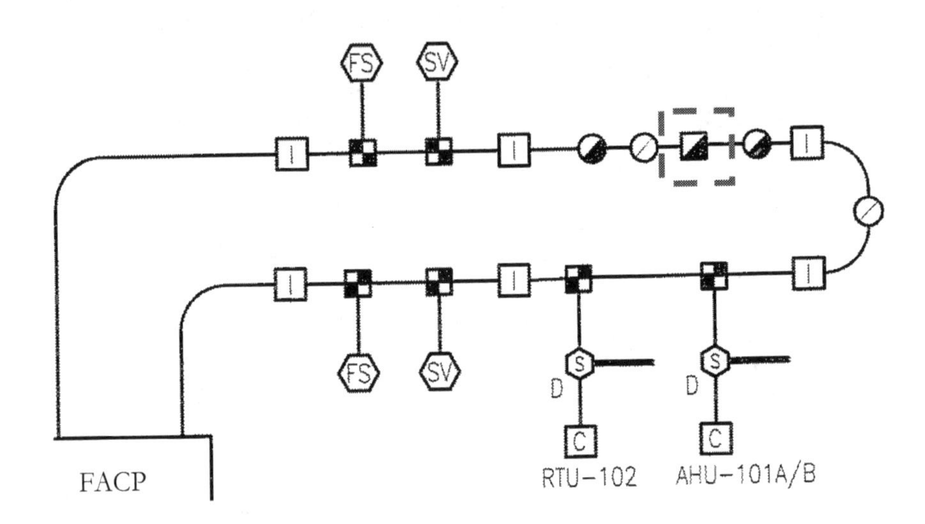


You need to understand that this loop includes more than one manual pull station. Observe the way I marked into a square box the manual pull station as an example of a device. Also note that I marked the manual pull station as an example of a device. Your project will have more than one pull station (the maximum number of smoke detectors <u>and</u> pull stations can reach 250). They will be indicated in the specific drawings and the way you'll find them represented is similar to this diagram. They can be on the same INITIATION loop or they can be on different initiation loops. Since one panel will have more that one loop each loop will likely include at least one pull station. These devices trigger the alarm when they are manually activated so they are not automatic devices but are integrated into the automated system. Basically this is an option designers will provide to the system to make it possible for humans to activate the system if they observe the escalation of a fire event into a protected area. These devices should be easily accessible. For this reason they should be installed near escape routes and close to the doors (no further than 900 mm (3'-00") at the 1200 mm height (4 ft ). They include a lever that can easily be pulled down when required. It is this lever that activates the system.

Manual pull station

**Manual pull station**

This drawing represents the main floor (grounding floor) of our model project. In the illustration you can see each area (room or corridors) where the smoke detectors (circle with red/white) are located. The manual pull stations are installed close to the entrance or exit doors not far than 900 mm (3'-00") at the 1200 mm height (4 ft). (See diagram on opposite page).
The arrows indicate the manual pull stations and their locations.

On this drawing (which represents the first floor of the building) we can identify about three manual pull stations. Later we will examine the connections type between these devices, other devices, and the way they complete the initiation loop.
If the FAS is a 2-stage FAS, then the manual pull station will have a micro-switch on it that will be activated in the way described in the previous chapter, "Single Stage & Two Stage Fire Alarm Systems."

Observe the message on the device's lever!

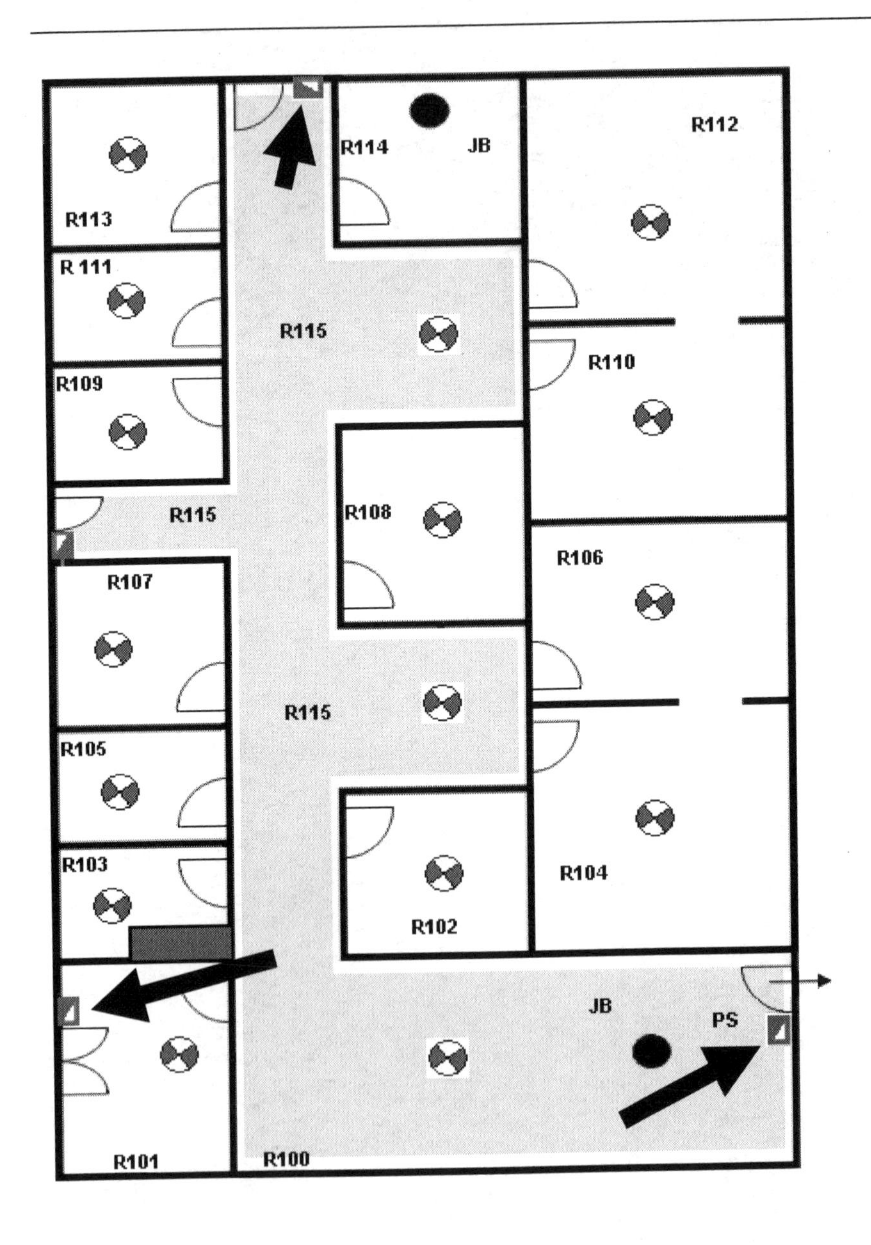


I want to show you the way manual pull stations are installed close to the doors. This will make you understand the scope of these devices and the role they play in a fire alarm system.

Please observe the diagram on the next page of a front door where manual pull stations are installed. Here there are two manual pull stations (one inside and the other outside the room).You can see how they are connected to the FAS. The magnetic lock is also part of the FAS and will be discussed on the next few pages. You need to understand the way manual pull stations are designed in order to be able to install them into the initiation circuit. These devices are not to be installed at just any existing door into the building.

Most of the time they are installed close to the escape doors (entrance and exit doors). If you look closely, you'll observe that these devices have a glass rod that needs to be broken in order for the lever underneath to be pulled down. If the glass rod is missing, it indicates that the pull station has been activated. This can be important information for the team that are managing the system either before or after a fire event.

Glass rod

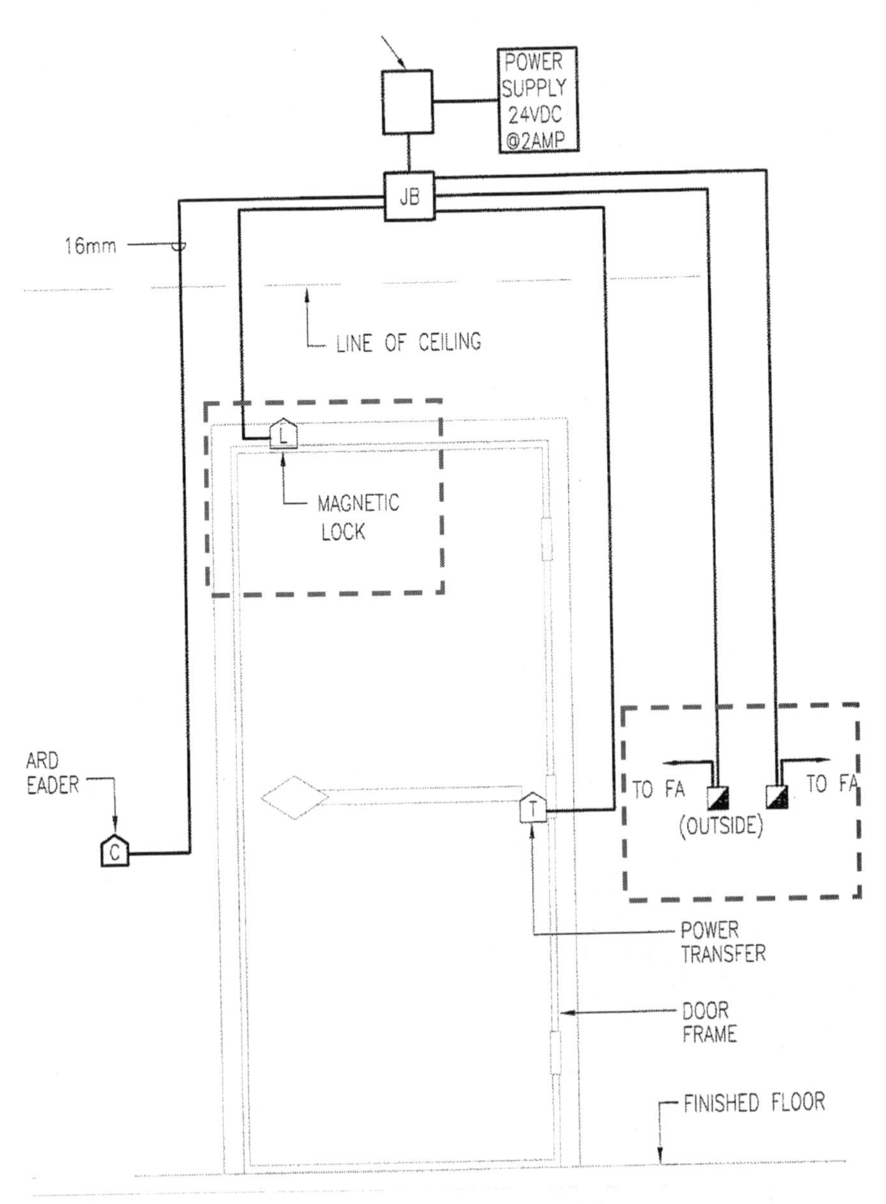


The following pages will show you how to terminate such devices. We will also clarify how pull stations are integrated into the loop.

- On the symbols table at the position seven (7) and eight (8) there are descriptions of the following symbols: ADDRESSABLE MONITOR/SUPERVISORY MODULE & ADDRESSABLE CONTROL MODULE.

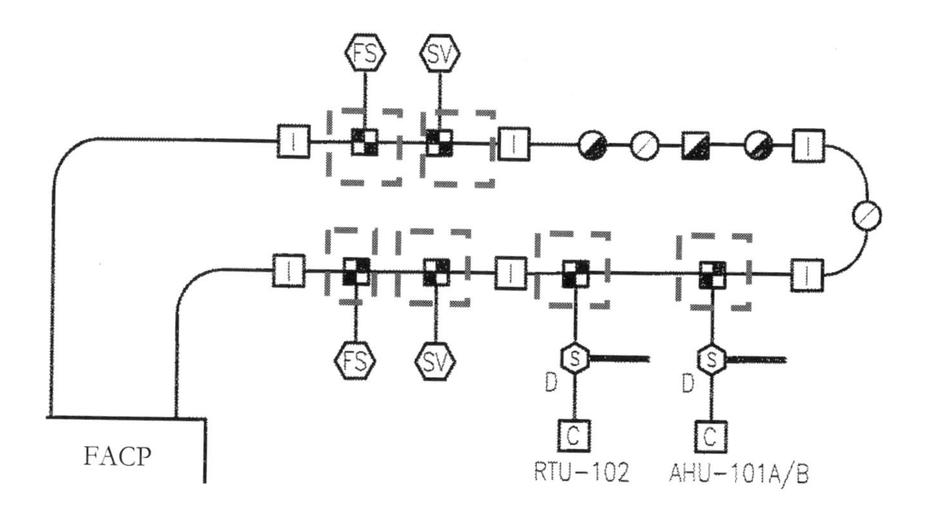


This loop does not include 6 (six) addressable modules only. The loop can have more or less modules to monitor, supervise and/or control other devices /systems.

Observe the way I marked the addressable modules into a square box to extenuate such devices.

They may be on the same INITIATION loop or be on a different initiation loop. These devices trigger the alarm when activated.

## ADDRESSABLE MONITOR/SUPERVISORY MODULE

These modules own an "ID" (such as a serial number) as an address and are introduced into the loop in the same way as the devices we've discussed in previous pages. Some of them do monitoring tasks or supervisory tasks for equipment or devices that are interconnected or interdependent with the fire alarm system.

The best example is the need for the FAS to monitor and supervise the main devices and equipment of the SPRINKLER SYSTEM.

(Most of you probably know what the sprinkler system is but some of you may not. It is a water pressurized pipe network that runs into the building. They are created to extinguish fires if they occur. Local devices will open special valves and will let water or gas as INERGEN to extinguish the fire. The pressurized water system is more traditional as the gas extinguishing system is relatively new. This new system works by having the gas "eat" the oxygen that is fueling the fire, effectively putting fire out.

Coming back to our scope of work:

The FAS does not control the sprinkler system but is doing the monitoring and the supervision of the valves, pressure, extinguishing and material flow. The sprinkler system is not triggered by the FAS but the FAS will indicate in real time the status of the main devices of the sprinkler system so that it is reliable and ready to perform in case of a fire event. These modules (monitor or supervisory modules) will have the capacity "to complete" the loop and at the same time to be interconnected with the sprinkler system. They will display the messages on the FACP or at the Annunciation panel. Some of the signals they transmit to the "brain" (FACP) will trigger the alarm of the FAS while other will only notify the system.

**Your notes here:**

-----------------------------------------------------------------

-----------------------------------------------------------------

-----------------------------------------------------------------

-----------------------------------------------------------------

-----------------------------------------------------------------

-----------------------------------------------------------------

-----------------------------------------------------------------

-----------------------------------------------------------------

## ADDRESSABLE CONTROL MODULE

These modules own an "ID" also which is like a serial number and address. It is introduced into the loop in the same way as the devices we've discussed in previous pages. These control some devices that are auxiliary to the fire alarm system. There are systems in most buildings that need to be controlled by the FAS. For example the access control into the building should automatically release all designated doors when a fire alarm system signal is triggered. No doors, other than those designated, should be locked by the access control system when a fire event occurs. In emergency situations these doors are released so that all people and animals may be safely evacuated. Some areas will require that doors remain closed or locked in order to avoid the spread of the fire to adjacent areas. System like HVAC (heating, ventilation and air conditioning) will have restrictions to bring fresh air into the areas where the fire is present or to take smoke from the contaminated areas. The elevators will also be under the FAS's control.

The FAS controls auxiliary devices such as:

- Door holder's release (Magnets on the wall to hold door)
- Magnetic Lock (locks the door for controlled access and should be released by the FSA during an alarm signal)
- Fan shutdown devices
- Elevator recall
- Pressurization fans

There are no differences between the addressable monitor modules and the addressable control modules. They can be the same type of device but their function (purpose or role) in the loop will determine whether they are doing a monitory or supervisory function or if they are performing a control function.

***"These modules own an "ID" which is like a serial number and address"***

All these devices will have an individual address that is unique and easy recognized by the fire alarm control panel (FACP). This individual address is going to be necessary when the system is ready for programming. Devices like detectors (SD or HD), manual pull stations (F), isolator modules (IM), control modules (CT1/2; CR) will have a unique address as intelligent devices part of the addressable fire alarm system. This unique address is like an ID (identification number) so when this device is working will be identified by the Faire Alarm Control Panel (FACP). In reality this specific unique address is transponder into a bare code like we see on any products. On its label this bare code I think is familiar to you:

Smoke detector (SD #101)

Smoke detector (SD #1xx)

Isolator Module (IM)

Control Relay (CR)

Detector (SD)

Pull Station (F)

Single Input/Output Module (CT1)

Later on our project these symbols and code bare will appear in our drawings. In fact will be a chapter that will indicate how to organize your-self in order to keep these bare codes available in separate binder for the programming task. I need you to appreciate the importance of these ID's for devices.

- At position nine (9) there is a description: DUCT SMOKE DETECTOR "SD"

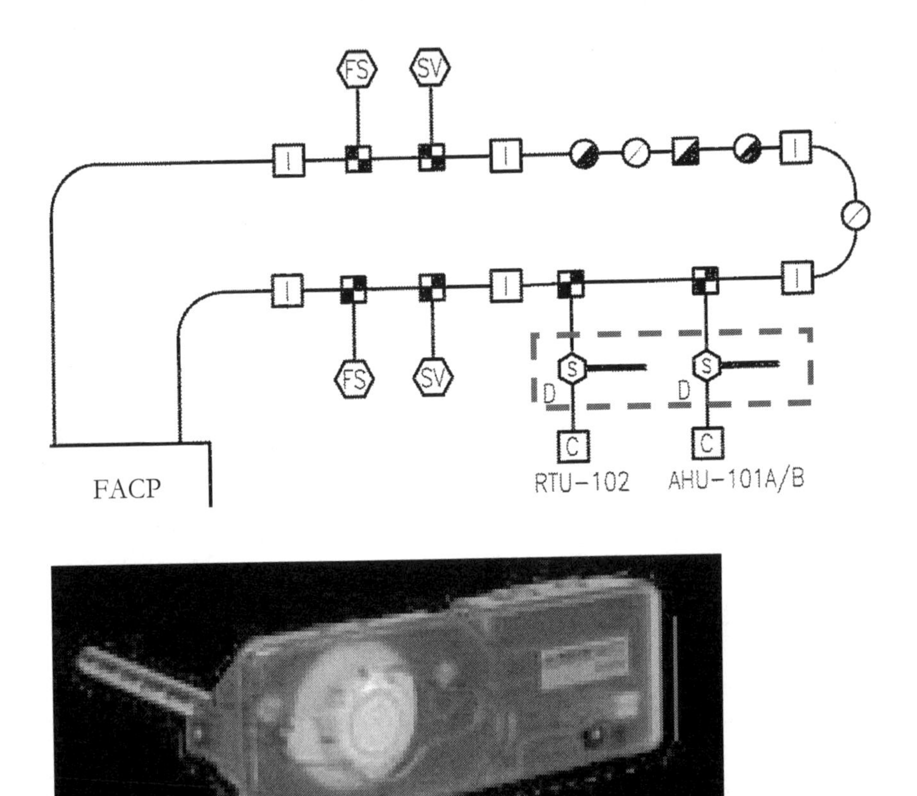


"SD"

The HVAC system is controlled system by the Fire Alarm System. The system shall have smoke duct detectors installed in order to identify situations where smoke is present into the duct. The fan that is running will then be shut down to avoid smoke being brought into the building. Often the location of the smoke duct detector will be indicated by the designer. The detector should be installed on the SUPPLY AIR DUCT. I will give a short and basic description of the typical duct system for any building so you'll be able to identify the location of the smoke duct detector. By definition the supply duct is the duct that distributes conditioned air (cooled, or heated, cleaned or humidified). The image below indicates the duct distribution system.

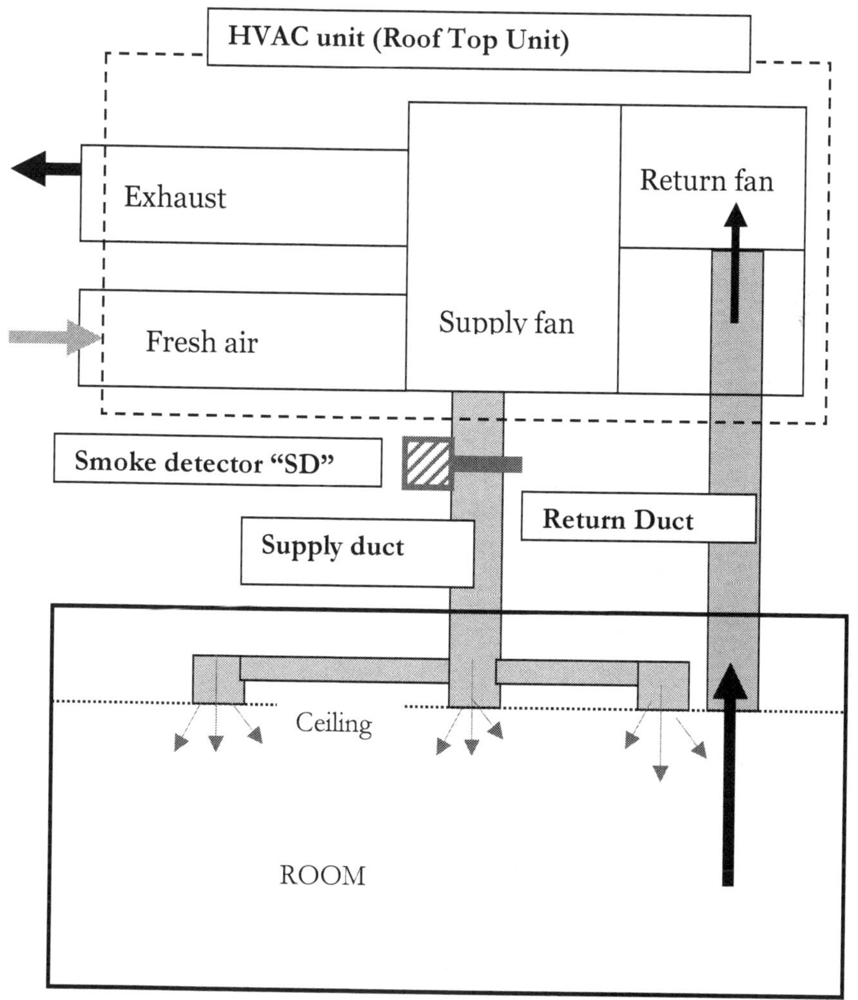

The smoke duct detector is located on the supply duct and will shutdown the supply fan located at the HVAC unit. In this way the smoke duct detector is controlling the HVAC system. In the event of an alarm signal on the loop from the SD, the FAS will shutdown the supply fan thus preventing smoke from contaminating the protected area. By definition the supply duct is the duct that distributes conditioned air (cooled, or heated, cleaned or humidified) .For the Inlet tube installation you should consider manufacturer recommendations.(holes to face the air flow to conduct smoke if exists into detector chamber)

TOP VIEW supply duct and the duct smoke detector installation.

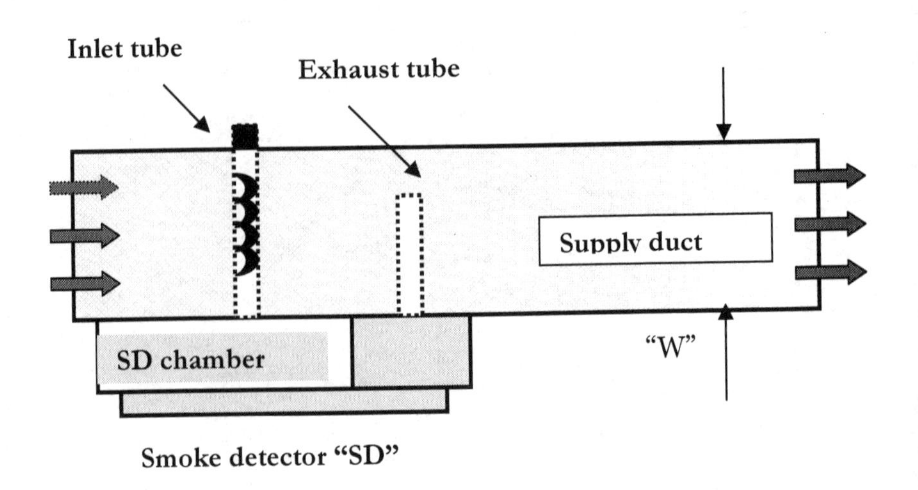

W = width duct.

Excess smoke will go back to the duct area through the exhaust tube. This device, as explained before, is an addressable one which owns an unique address recognized by the system's software due to programming.

Let illustrate a side view of the duct to understand how this SD is installed on the duct and to observe what information will display

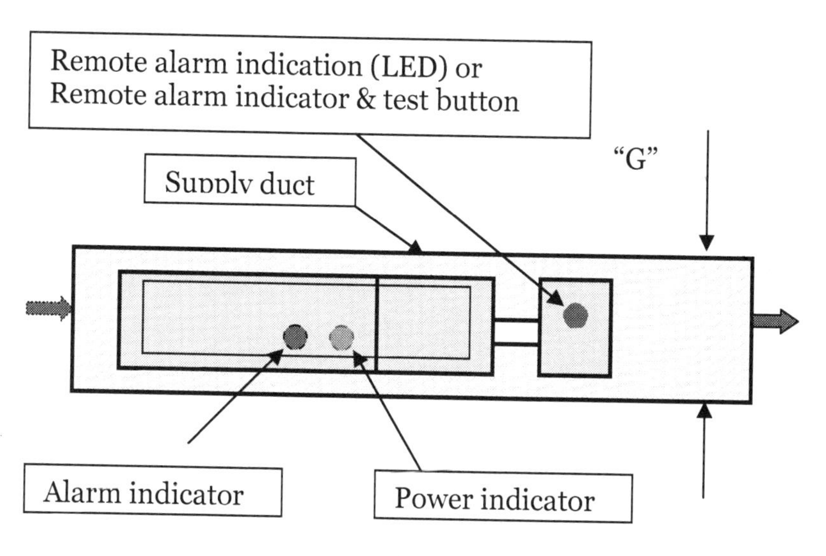


G= thickens of duct

In installations where the duct smoke detector's controls and indicators are hidden from view for the spaces reasons, a remote test station or an LED indicator can be connected to the detector to provide these functions.

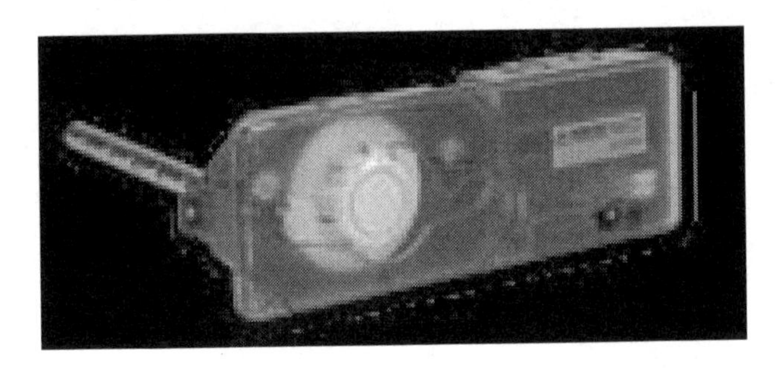

There is an alarm relay output also available on these smoke duct detectors. Most will provide at least a dry contact relay as normally –open and normally closed contacts. With this available there will be no problems controlling any auxiliary devices or equipment like fan shutters (doors or windows covers) or fire dampers. A wiring diagram for such devices will be shown on next few pages. The device is easily connected to the loop and uses two wires cable (IN & OUT). Coming back to the original loop diagram for the portion indicated in the square box, I need to make the following comments:

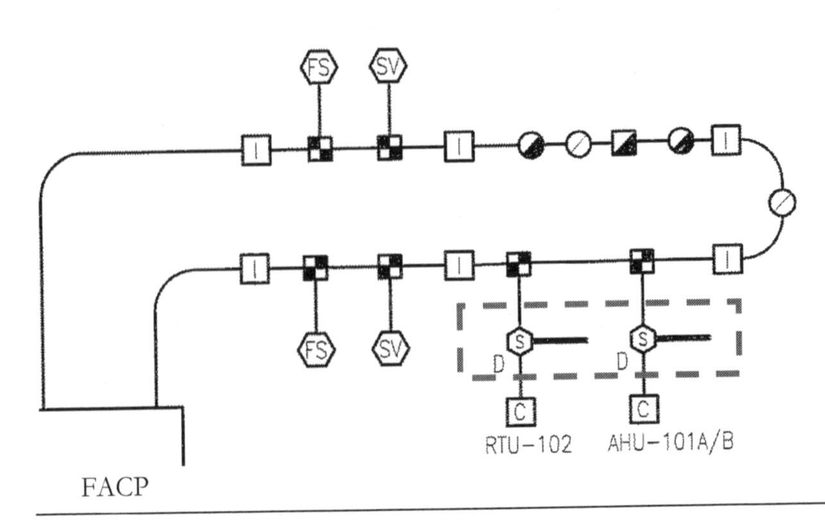

RTU -102 stands for Roof Top Unit #102

AHU-101 A/B stands for Air Handling Unit # 101 A & B

C represents the controls logic diagram of RTU and AHU

In other words, we can control with such these devices (smoke duct detectors) equipment like RTU and AHU since the alarm signal exists

**Your notes here:**

---------------------------------------------------------------------

---------------------------------------------------------------------

---------------------------------------------------------------------

---------------------------------------------------------------------

---------------------------------------------------------------------

---------------------------------------------------------------------

---------------------------------------------------------------------

---------------------------------------------------------------------

---------------------------------------------------------------------

---------------------------------------------------------------------

---------------------------------------------------------------------

---------------------------------------------------------------------

---------------------------------------------------------------------

---------------------------------------------------------------------

- Position eleven (11) shows the: SUPERVISED VALVE

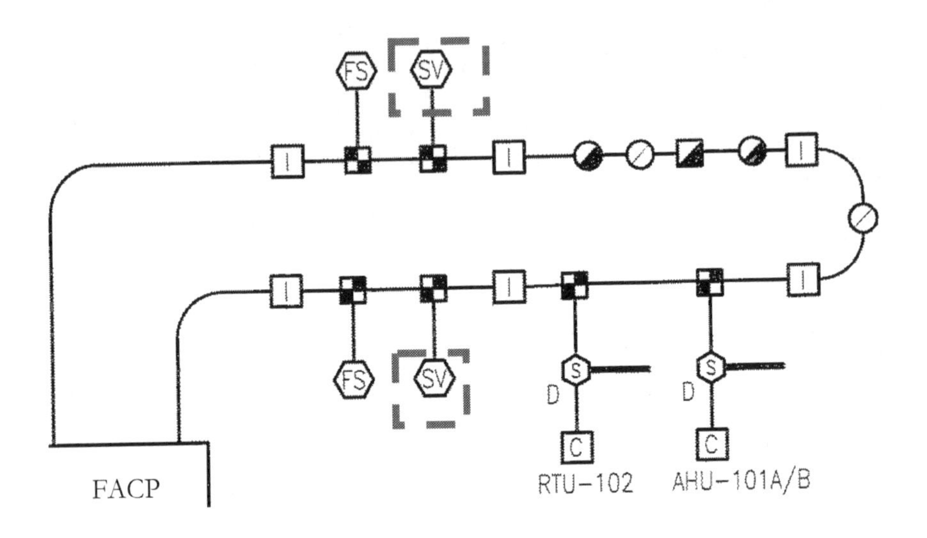


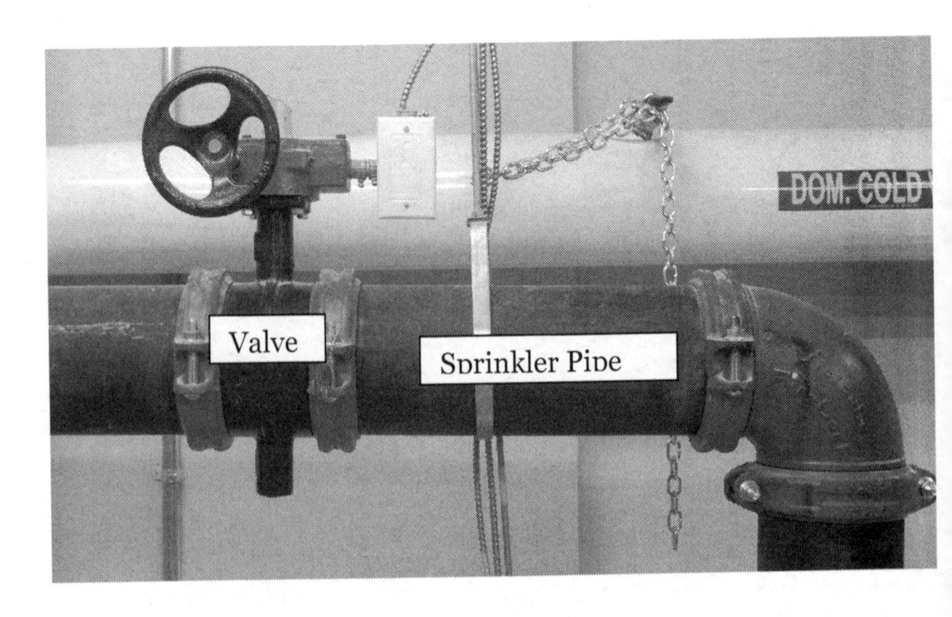

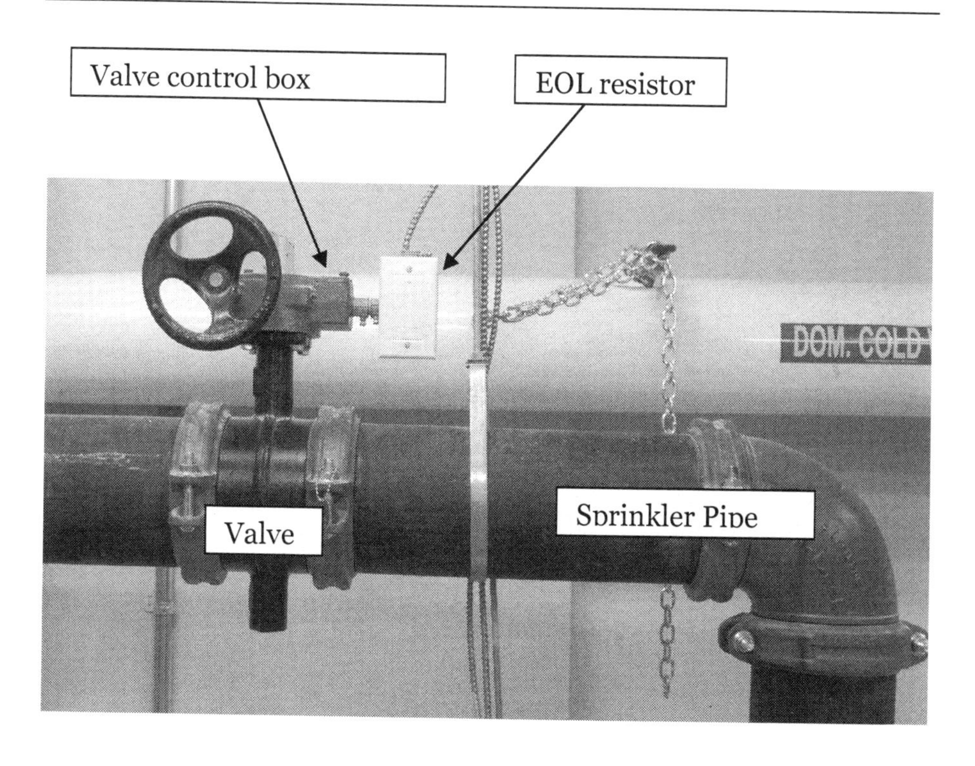


This is the sprinkler system. The valve needs to be supervised. This circuit indicates the status of the valve (VALVE OPEN or VALVE CLOSED). Let us assume that the valve should stay OPEN during the normal operation of the sprinkler system. This valve should remain in the OPEN position all the time so the sprinkler system is ready to work and respond properly when required during any fire emergency.

Any change of status will be indicated on the Fire Alarm Control Panel (FACP).

This is a valve monitoring/supervision by the Fire alarm system.

Let zoom in on the valve area ...........

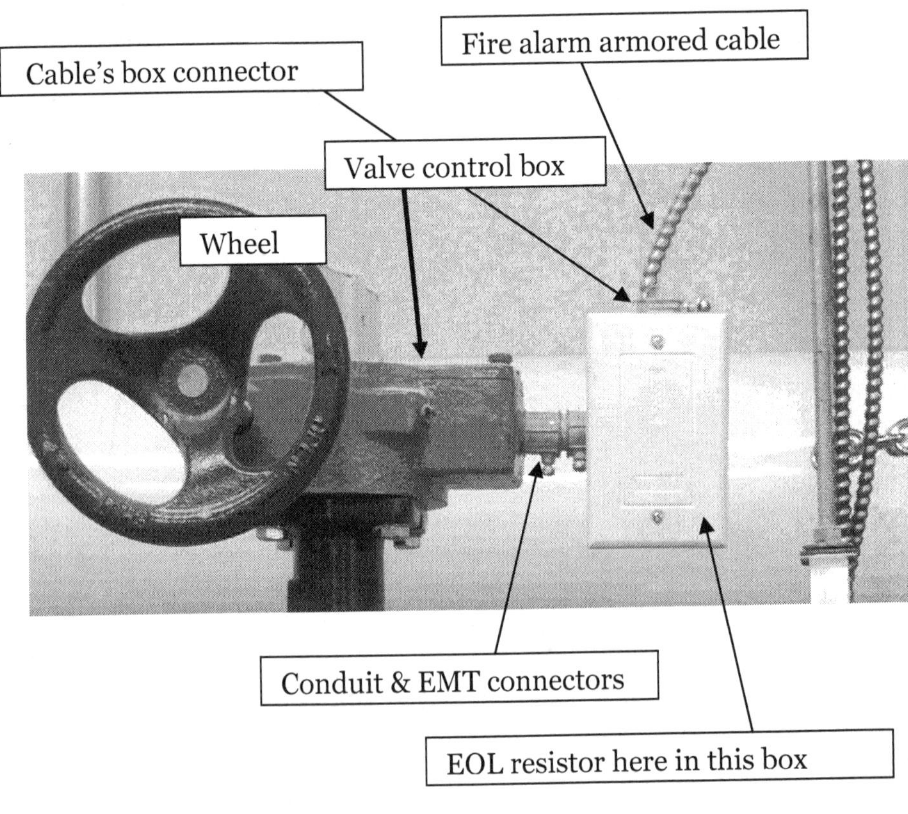


Your notes here:

........................................................................................................

........................................................................................................

........................................................................................................

........................................................................................................

........................................................................................................

........................................................................................................

........................................................................................................

The way the wiring is terminated at this location of the valve control box is displayed below:

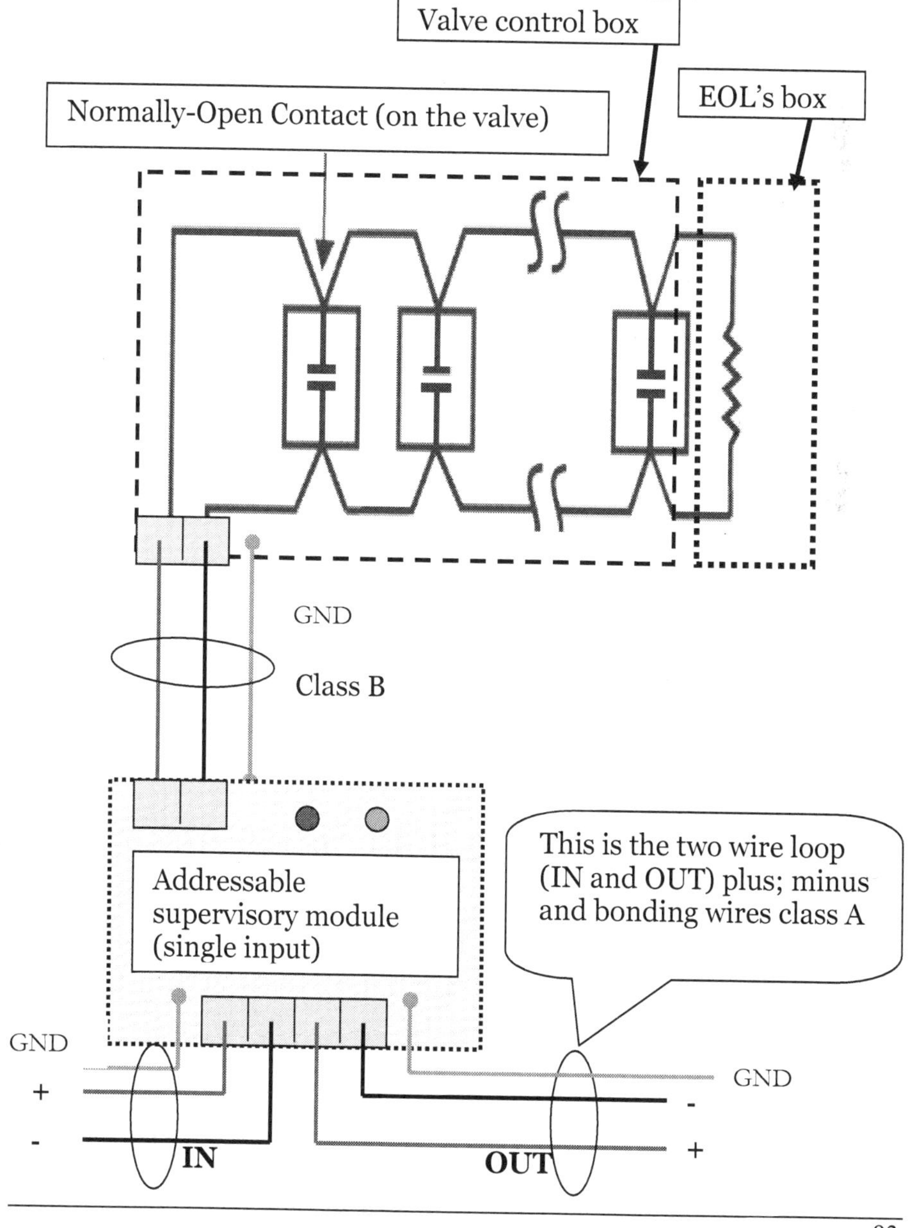

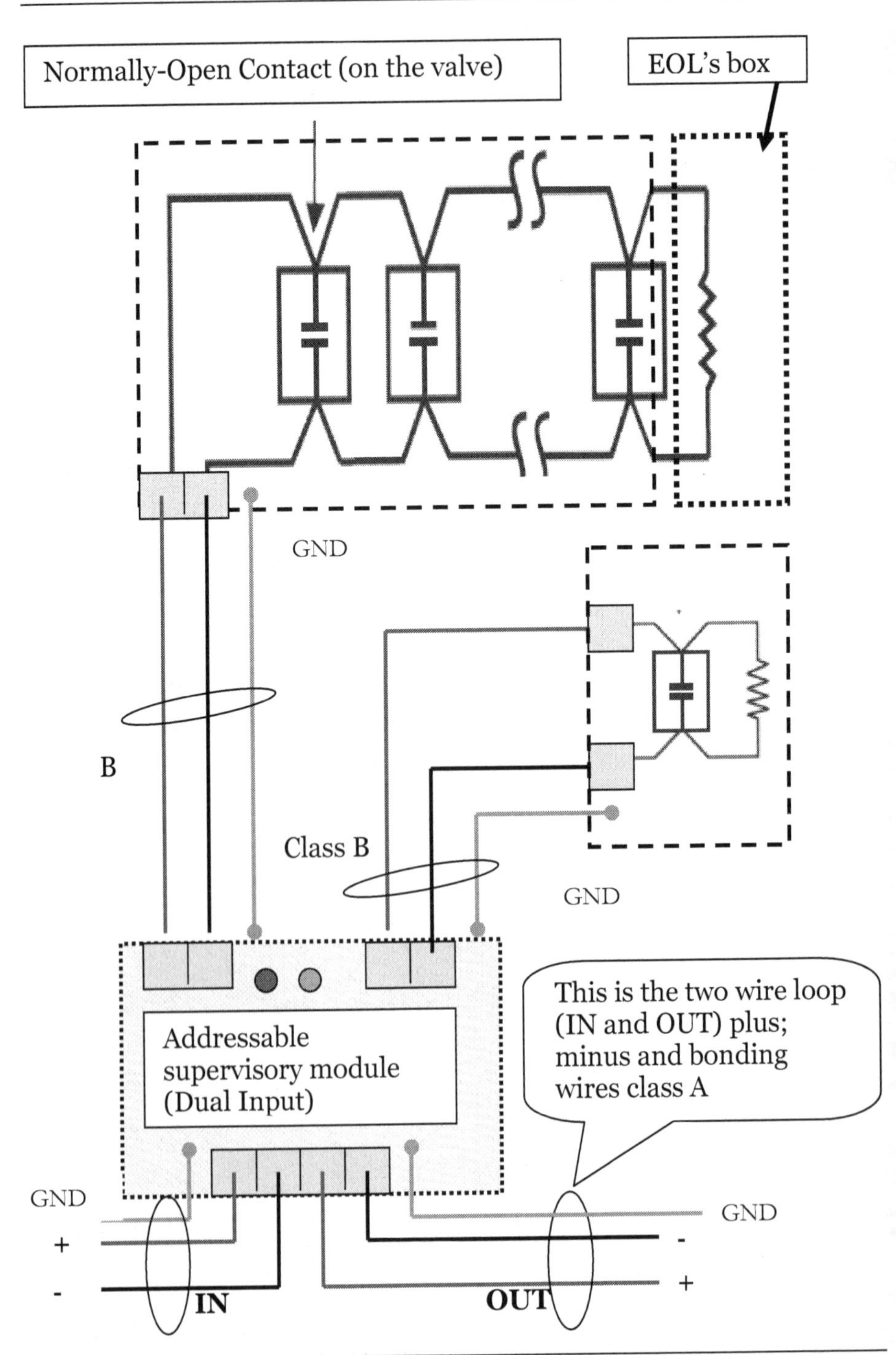
Normally-Open Contact (on the valve)
EOL's box
GND
B
Class B
GND
Addressable supervisory module (Dual Input)
This is the two wire loop (IN and OUT) plus; minus and bonding wires class A
GND
+
-
IN
OUT
GND
-
+

As you can see there are available two types of addressable modules to be integrated on the loop. These are single and dual input/output modules (I/O).The first example is referring to the Single input /output module since the second is a dual input /output module. Both modules are considered intelligent analog addressable devices. The single I/O module is able to connect the loop to one class B circuit. The function of such a circuit could be monitoring, supervisory or alarm with dry contact on the initiating circuit. The dual I/O module is able to connect the loop at two class B circuits. The function of such circuits is for monitoring or supervisory or alarm with dry contact on the initiating circuit. For example when using a dual I/O module you are able to supervise a valve and to monitor the pressure of the water on the sprinkler system. The industry is currently using different types of devices. I worked with many types of devices manufactured by the GE Security (Edwards's system). Available addressable modules from this manufacturer are SIGA-CT1 and SIGA –CT2. The CT1 is a single I/O module since CT2 is the dual I/O module.

As you'll see, the way these modules work is:

The module is able to convert the analog signal from the initiating devices (normally-open contacts or normally–closed contacts) to a digital signal. This digital signal is introduced into the loop and is interpreted by the fire alarm control panel. The role of the end-of–line resistor is to supervise the circuit B integrity status. This will guarantee that the analog signal from the initiating device travels through the wires to the addressable module where it will be converted into a digital signal. If something happens to compromise the physical integrity of the B type circuit, a trouble signal will appear at the FACP. If any accidental grounding of the line occurs, this will be indicated as well.

SIGA –CT 2

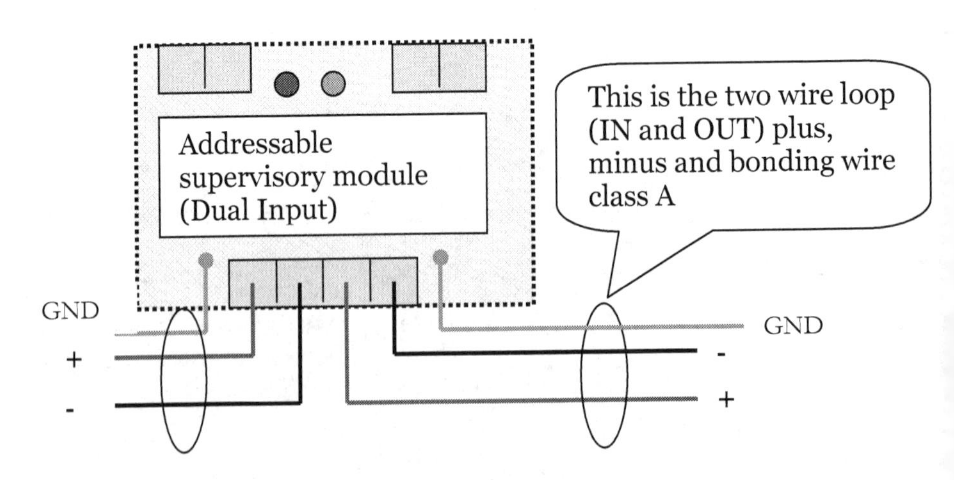


- The **red LED** = ON is indicating the Active ALARM since
- The **green LED** =ON is indicating the NORMAL status of the system.

This single or dual I/O module will fit into a 2 ½ "single-gang metallic box. This device is provided with a suitable cover to fit on the single- gang box as been indicated. The terminal block for class B and Class A INI circuits will permit termination at the #12 and # 18 AWG wires.

#12 AWG = 2.5 mm Sq

#18 AWG = 0.75 mm Sq

This device should be installed in a visible location and be accessible for the purpose of maintenance. You need to consult the manufacturer's specifications because they often indicate the size & type of box required to permit the installation of such a device.

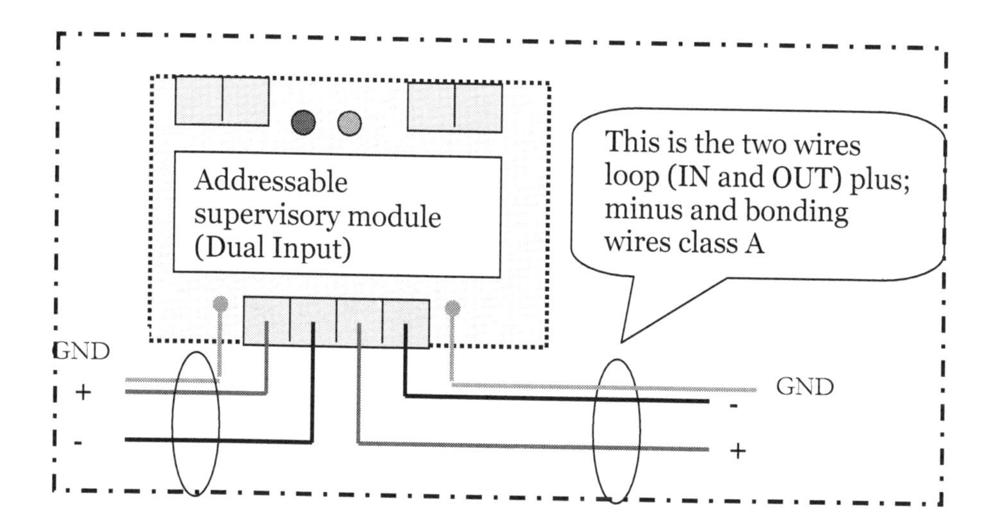

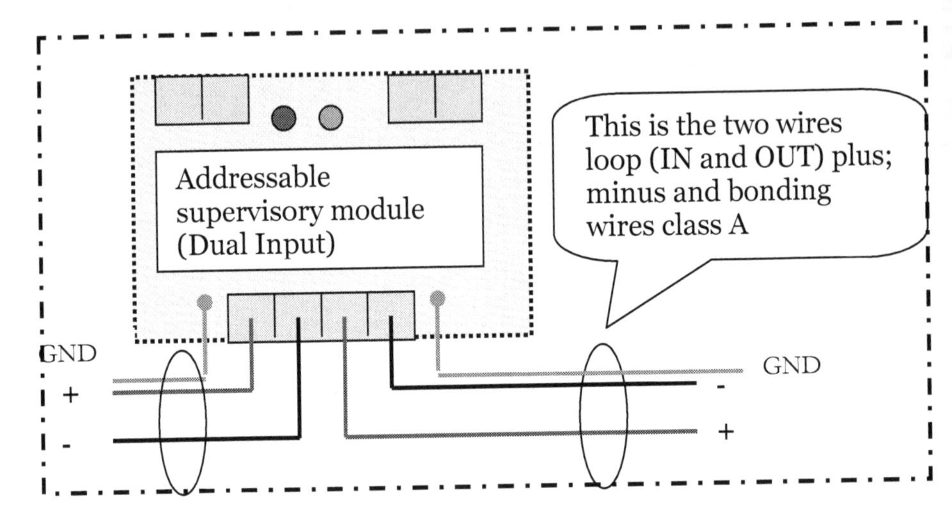


Supervisory Module (addressable) : see its "place" on loop diagram...

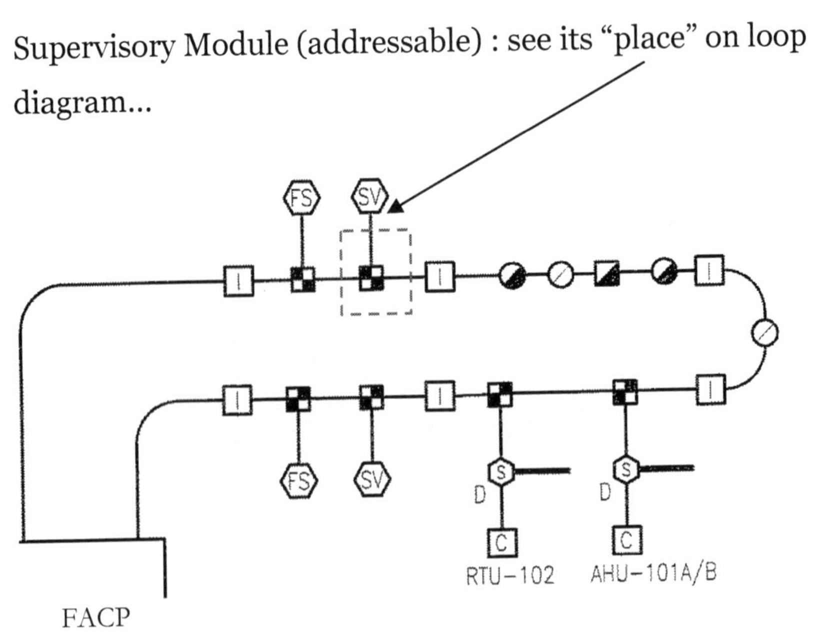


A 2 1/2" deep single gang electrical box is ok for module installation.

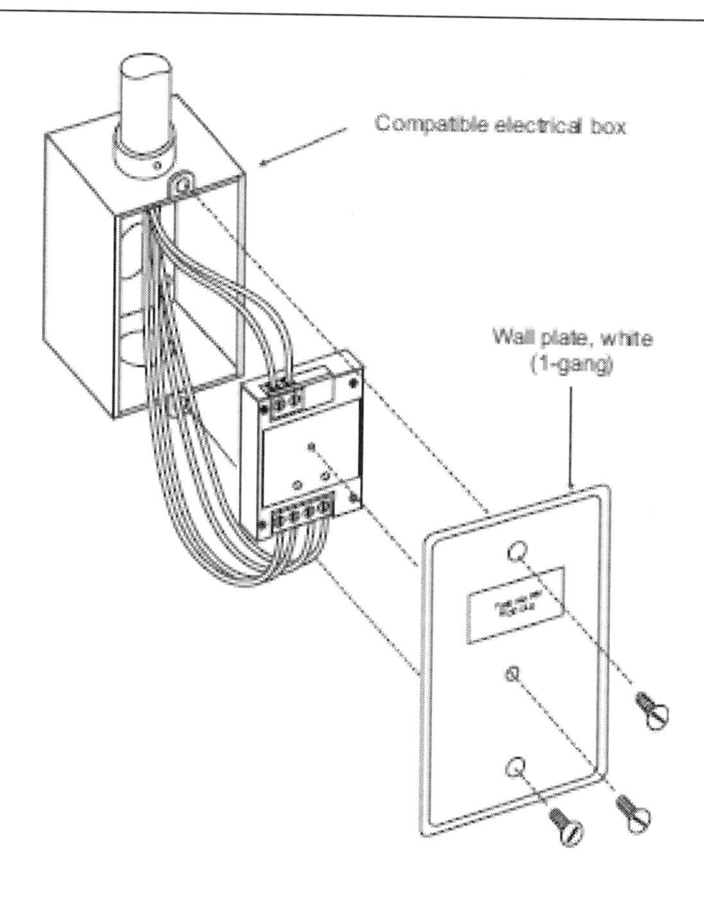


Here you have an idea how and where the control module should be installed

This is the diagram you should use when considering installation of an addressable module. Obviously you'll install a support for this box unless there is a surface mounting detail on the wall or existing support. Observe the conduit's connector and the spare length of wire inside the box. These are items you need to consider when delivering this product, as you have to ensure easy maintenance, high quality work and a functional, reliable system. The face plate of this device is designed in such a manner that the green LED & red LED will be visible when flashing. Green flashing means the system is normal whilst red indicates an alarm signal. The end–of–line resistor, which supervises the line's integrity, will be located into a separate box or even into the valve box.

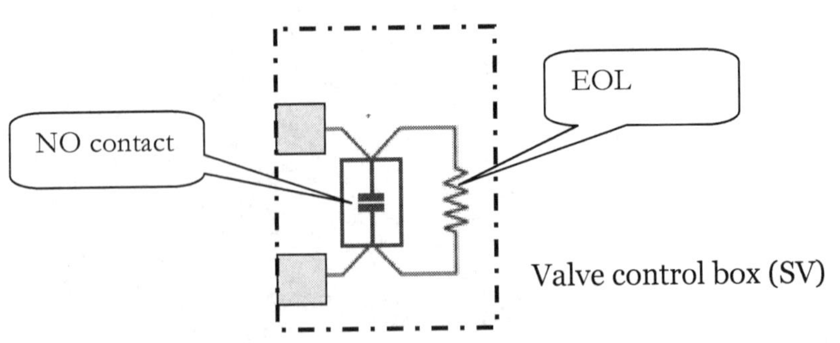


This device should be installed in an accessible location and be easy to reach for maintenance. A single-gang electrical box set 2.5 " deep is required to properly install the EOL. A faceplate will cover the box. In order to understand this better, have a look at the following details:

Let zoom the area..........

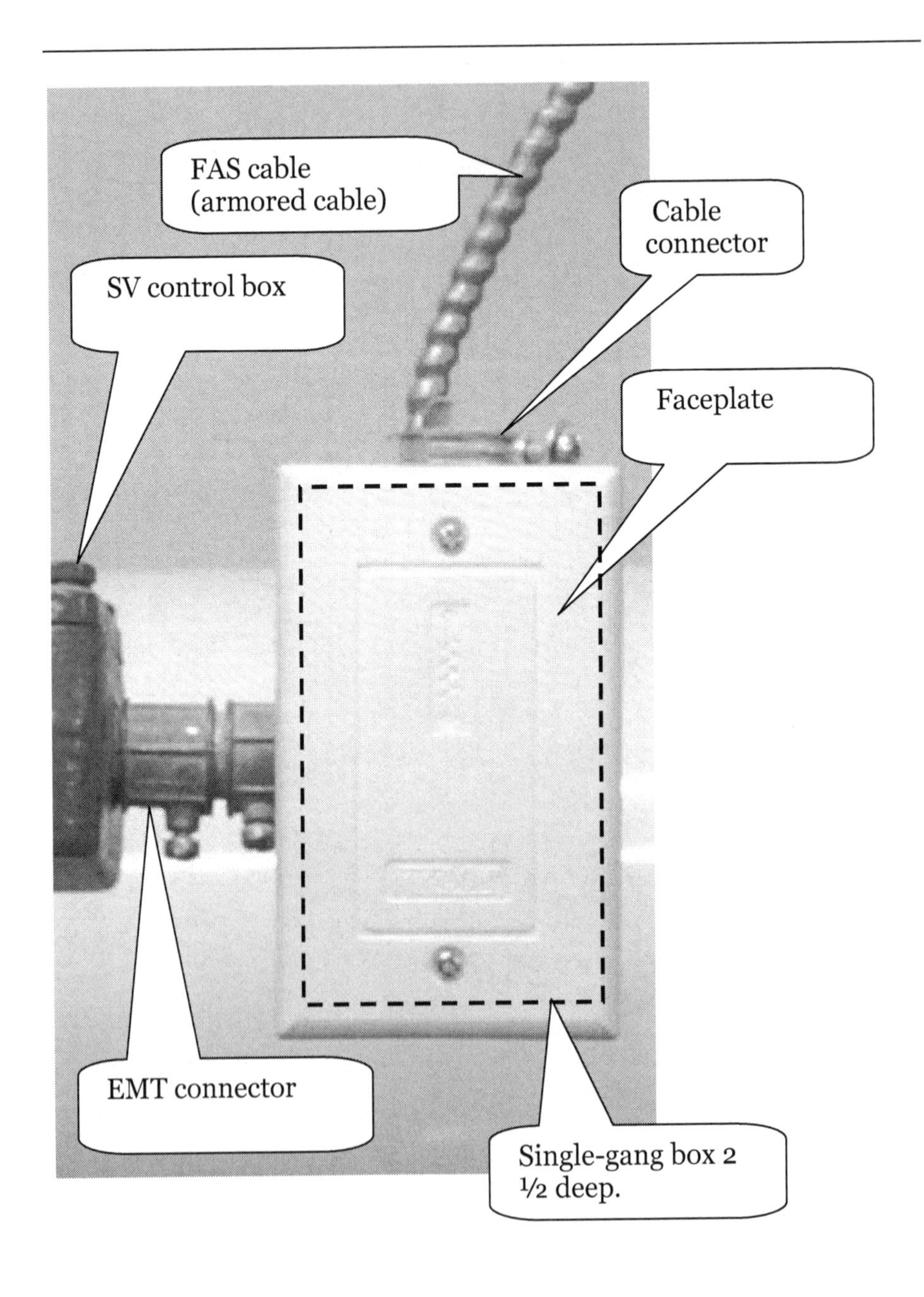
FAS cable
(armored cable)
Cable
connector
SV control box
Faceplate
EMT connector
Single-gang box 2
½ deep.

Same details but the SV is on a vertical position. Observe the box for EOL resistor!

Your notes here:

..................................................................................................

..................................................................................................

..................................................................................................

..................................................................................................

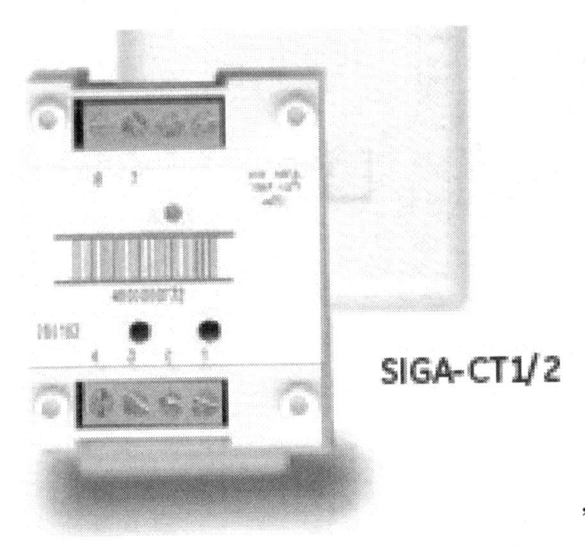


"LIVE" Device SIGA-CT2

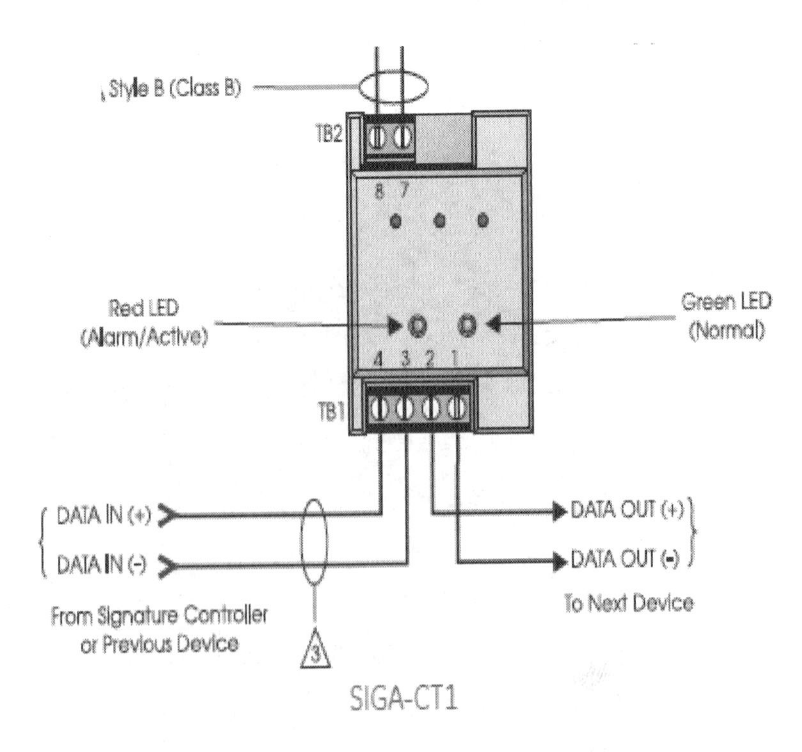


Diagrammatic SIGA –CT 1

- Position ten(10) is indicating description like: FLOW SWITCH

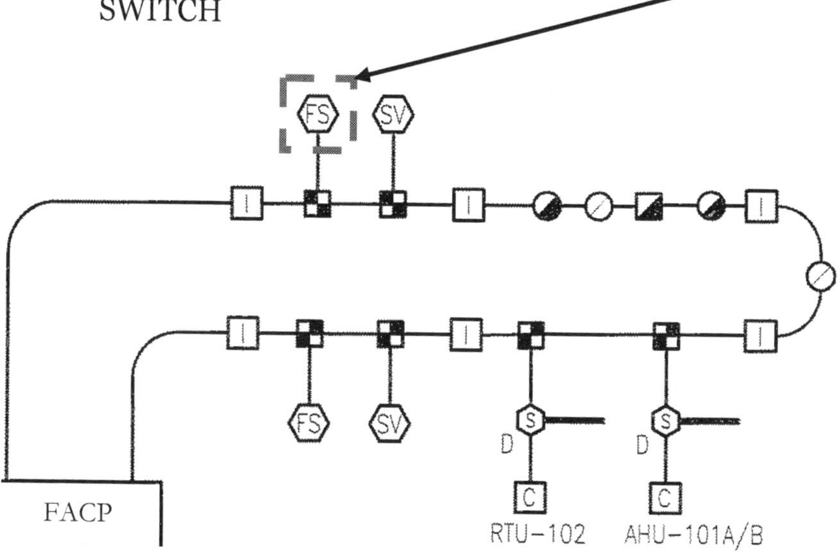


The flow meter is responsible for providing the information to the Fire Alarm System that the sprinkler system is working appropriately. The main components and elements of the sprinkler system such as valves, pressure into the sprinkler system and the flow detections pertaining to the extinguishing fluid of the system are monitored or supervised. This is the guarantee that the fire alarm system and the sprinkler system will work properly when required. On the market you'll find different types of flow meters but our example will show some that I've encountered many times during installations. The diagram shows a flow switch installed on the vertical sprinkler pipe. Observe the fire alarm cable entering the flow-switch on the back side. The EOL resistor is installed inside the flow-switch box. This is when a class B circuit is installed. The switches are activated when a flow occurs through the device. This information will generate a signal to the FACP. This means any unexpected flow- for example a broken pipe- in the sprinkler system is being detected and the information will appear and be available for maintenance personnel when they arrive to repair the system.

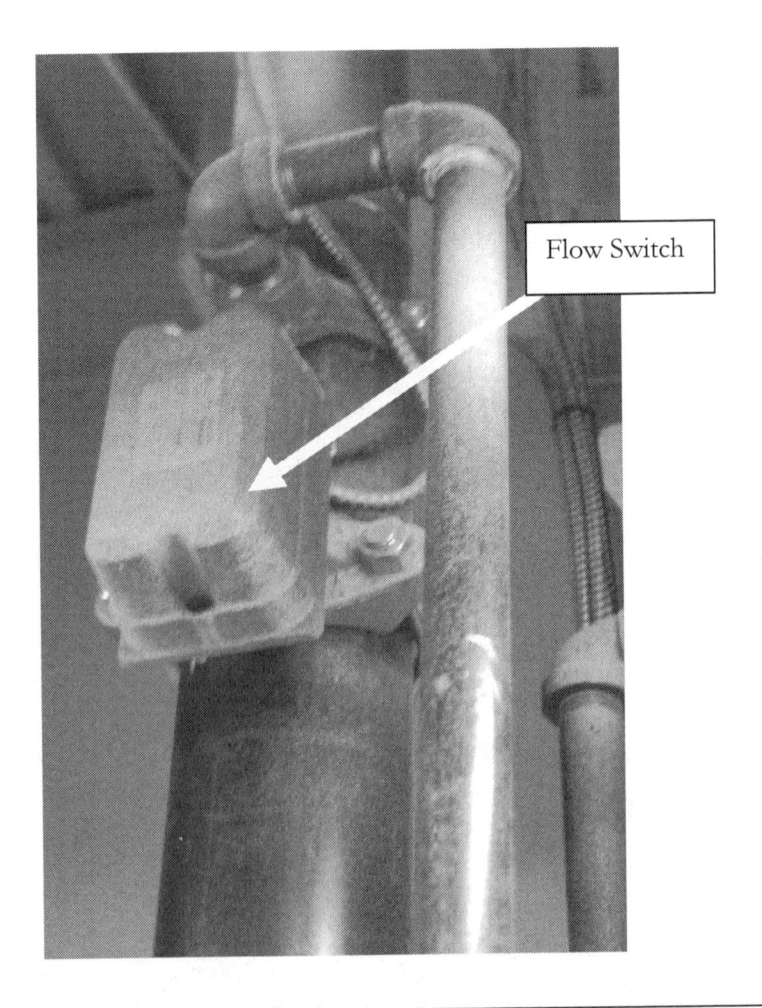

This device is connected to the loop using the same type of module we presented before at the supervised valve. The device is either SIGA-CT 1 or SIGA –CT2.

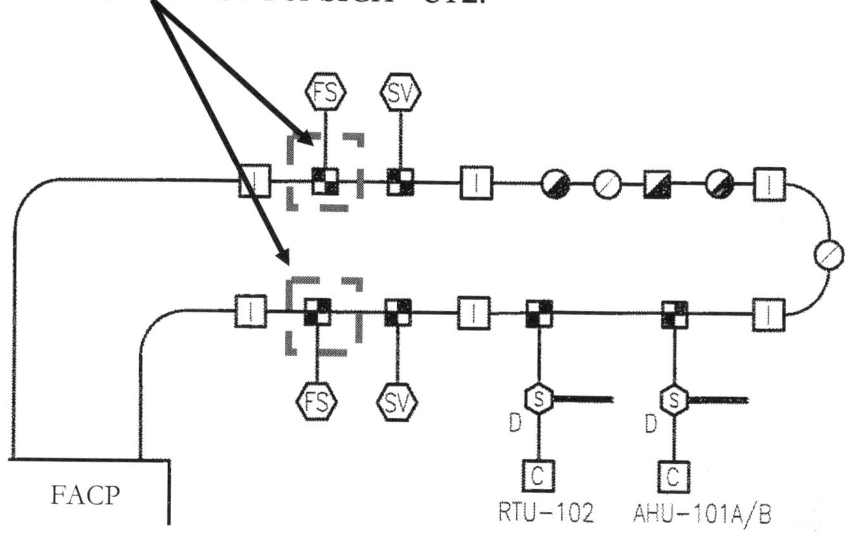


Let me remind you what a SIGA-CT2 is:

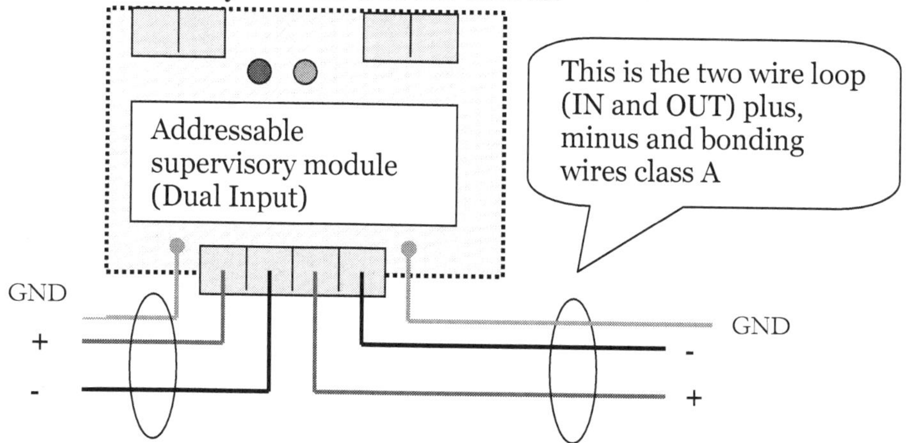


- The **red LED** = ON indicates an Active ALARM since
- The **green LED** =ON indicates the NORMAL status of the system.

This single or dual I/O module will fit into a 2 ½ "single-gang metallic box. The device is provided with a suitable cover to fit on the single- gang box as indicated. The terminal block for class B and class A INI circuits will permit termination with the #12 or # 18 AWG wires.

#12 AWG = 2.5 mm Sq

#18 AWG = 0.75 mm Sq

This device should be installed in a visible and accessible location for maintenance purposes. You need to consult the manufacturer's specifications because most of the time they indicate the size & type of box required for the installation of such devices.

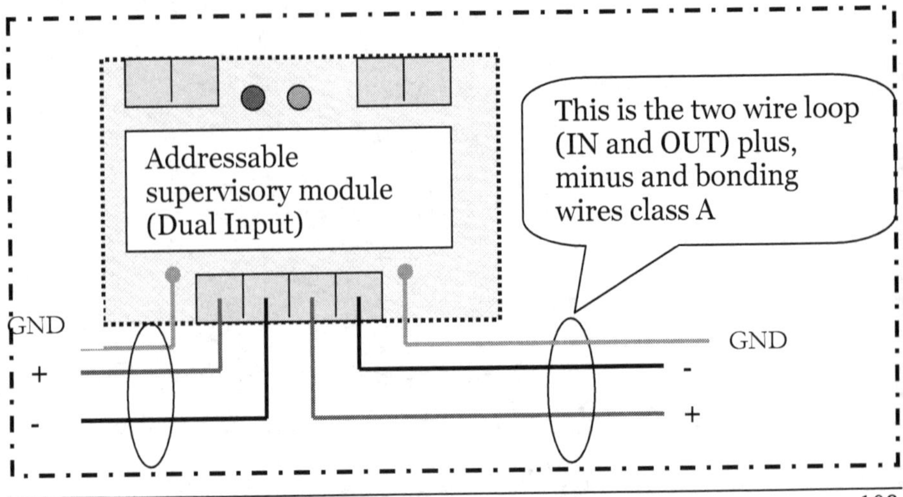

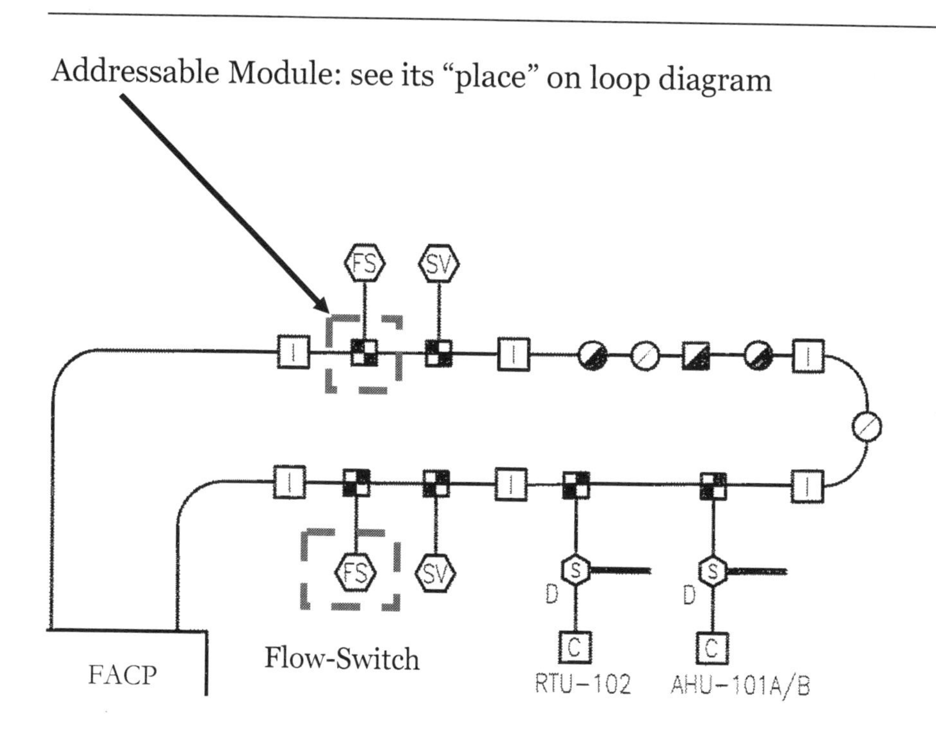


**Your notes here:**

------------------------------------------------------------------------

------------------------------------------------------------------------

------------------------------------------------------------------------

------------------------------------------------------------------------

------------------------------------------------------------------------

------------------------------------------------------------------------

------------------------------------------------------------------------

------------------------------------------------------------------------

The wiring will be terminated in the same manner as previously indicated at the supervised valve. The EOL resistor could be installed in the flow-switch box or into a separate box as indicated at the supervised valve.

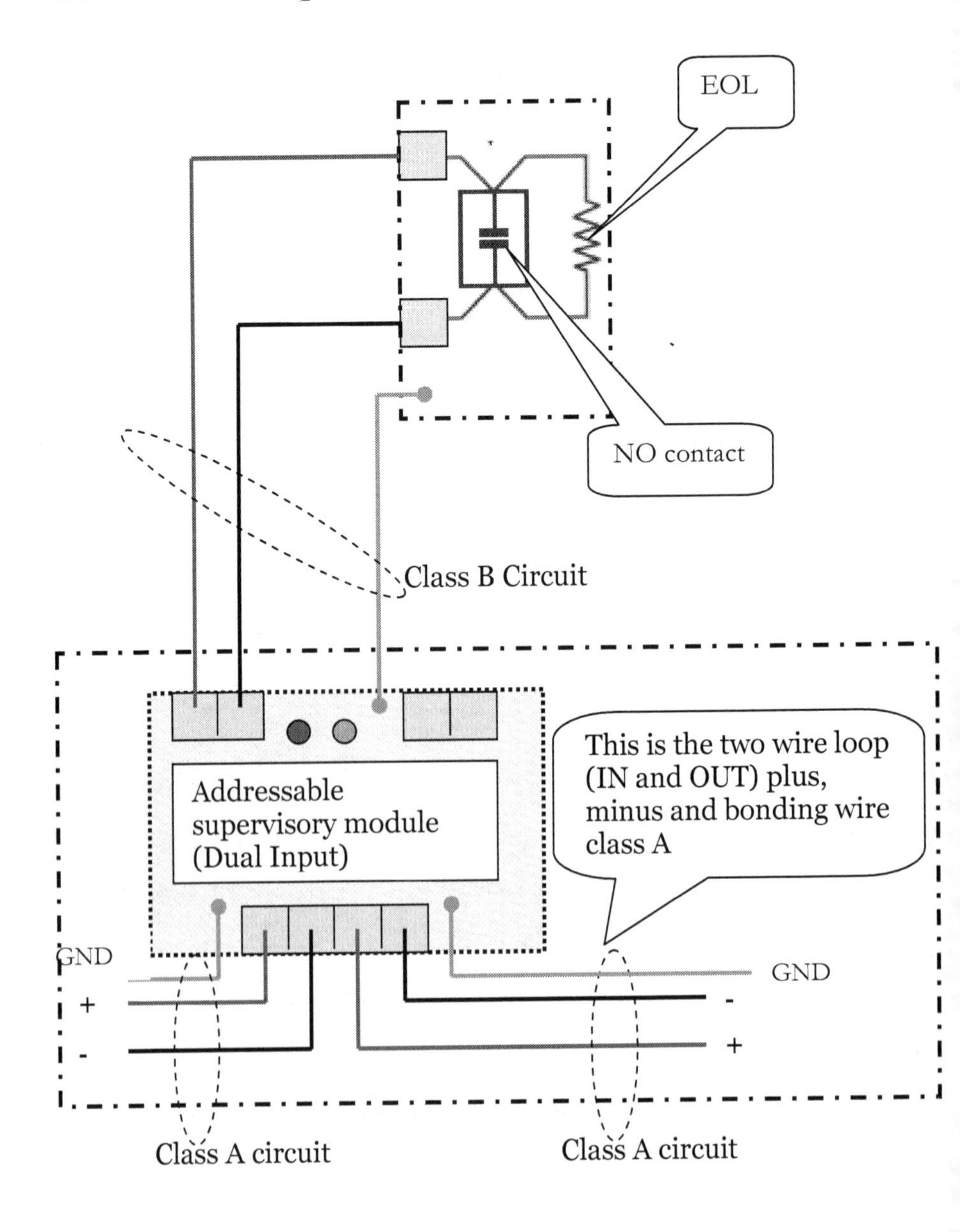

- Position twelve (12) showing a description like: ISOLATOR MODULE(IM)

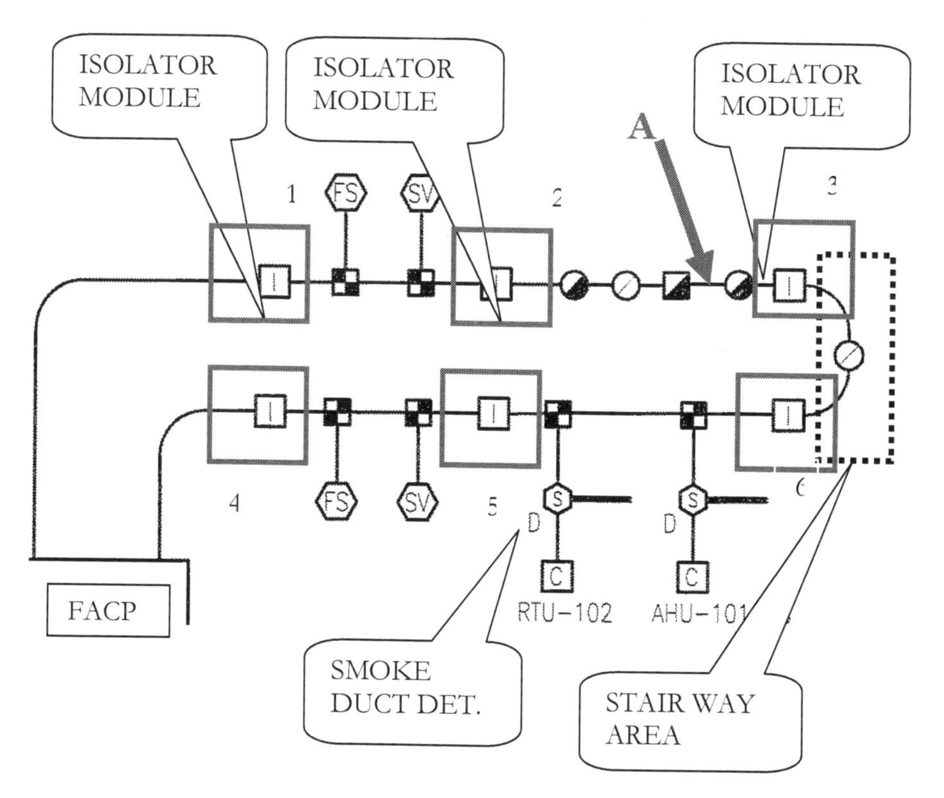


This device (IM) is able to isolate the area that has defects and make power available to the system via a different pathway. They are installed in areas that are critical during a fire or emergency. For example, the stairs are the means of evacuation so it is imperative that stairwells do not become contaminated with smoke. Having a detector located in the stairwell is mandatory.

In order to maximize the availability of the detector in the stairwell area, two devices indicated like [I] are found outside the stairs at the "IN" side and "OUT" side of the loop (I/O). Let's observe how this kind of device works:

Let's say there is a defect in the loop. For example, a line is damaged, broken or grounded at point **A**. This will compel isolator#2 and isolator #3 TO OPEN THE LOOP thus disconnecting the power for the devices located between IM #2 and IM #3.In doing this, the defect from the loop is eliminated but the detector located in the stairway continues to work.

The bottom line is that the devices between the isolators will be protected and their availability will be ensured in case the line suffers any damage during or before a fire event. There are also detectors located in the elevator shaft (smoke detectors) and elevator pit (heat detectors). They are protected in the same way since elevators need to respond in specific ways when the fire alarm system is activated. They should go to designated floor locations (homing) and be prevented from stopping between floors. In order to make it easier to understand the installation process for the ISOLATOR MODULES in the stairwell area, let's make some sketches. Most stairwells will have a smoke detector installed at the top level of the stairs, on the ceiling. In our diagram the detector is located between isolator module # 3 and isolator module #6. The way in which I represented the portion of the loop on the front view shows the smoke detector inside the stairwell since the isolator module IM #3 & IM #6 is located outside on the wall. The wiring of the circuit is protected inside EMT (electrical metallic tubing). There should be no confusion when looking at the next details. There are two details provided: the front view and top view. You'll also observe a manual pull station close to the door, outside the stairway area. This manual pull station is part of the INI (initiation) circuit which is connected to the INI loop. The symbol for the manual pull station is: "**F**"

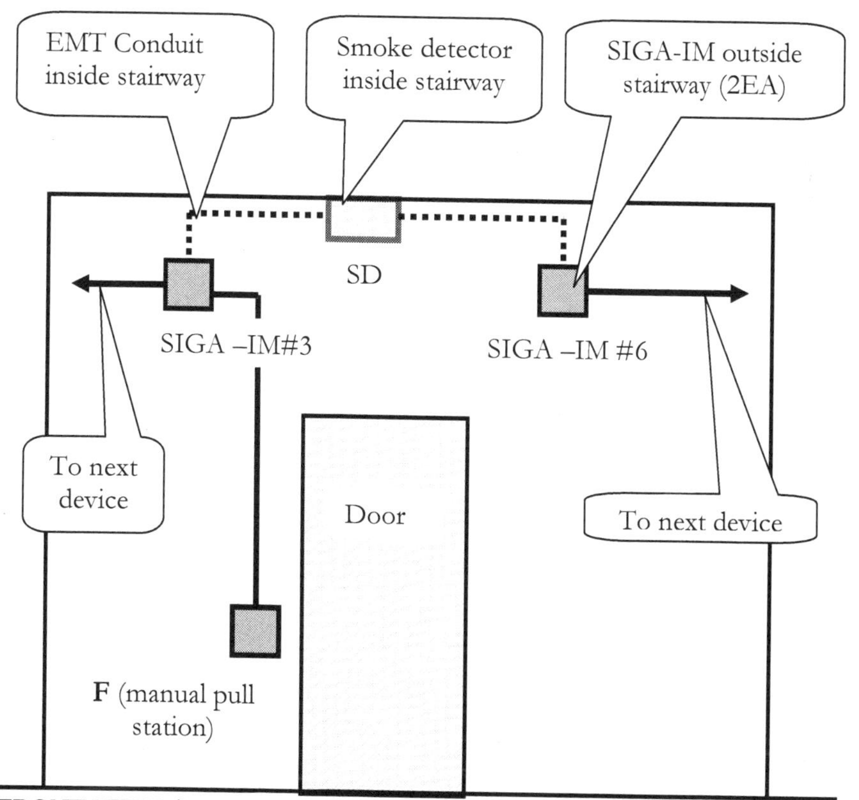


FRONT VIEW /LAST FLOOR smoke detector & Isolation Module –IM (stairway detail)

The SIGA-IM Isolator Module is part of GE Security's Signature Series system. I've selected this to use as an example. Check their site for clarification and information about this manufacturer. TOP VIEW STAIRWAY

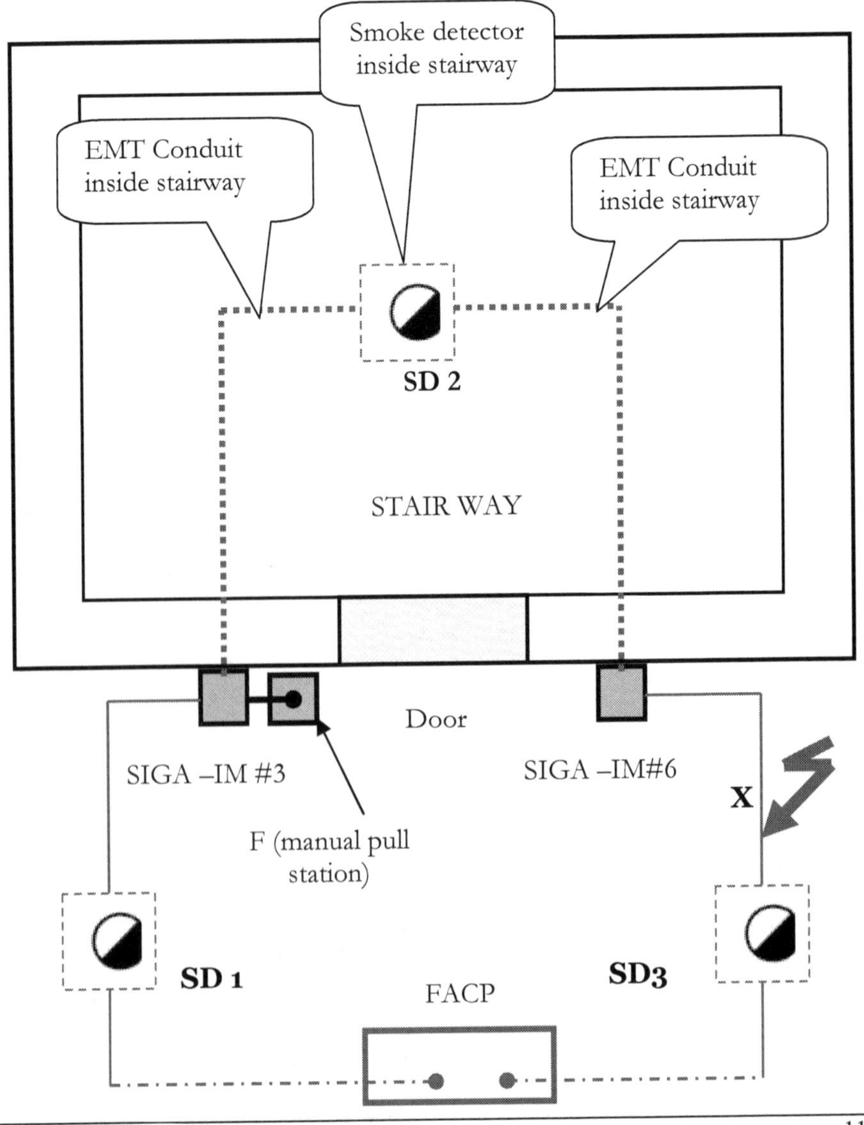

As previously mentioned, this device is able to isolate the "defective area" and keep the system running. There is a sequence to how the IM isolates the defective area. First of all, if the short is on the line, let's say at point "**X**" ALL isolators (IM #3 and IM#6) will open within a few milliseconds.

1. IM # 3 –OPEN
2. IM# 6 - OPEN

After that, starting from one side of the loop, the isolator will close to provide power to the next isolator.

3. IM # 3 – CLOSE

When the closest isolator from the short (IM#6) closes and the fault still exists, within seconds it will reopen. After this there will be continuous verification to see if the fault has been eliminated. If not it will keep the line open, but the rest of detectors will work.

4. IM # 6 - OPEN

The main positive from this is:

In the stairwell the detector is available to control the area even the though the line is defective. Furthermore, the system is available to provide information about the situation in critical areas like stairwells before or during an event in which the circuit has been damaged.

5. DS # 1 & DS #2 – WORKING
6. DS #3 – NOT WORKING
7. DS# 1;2;3 –WORKING

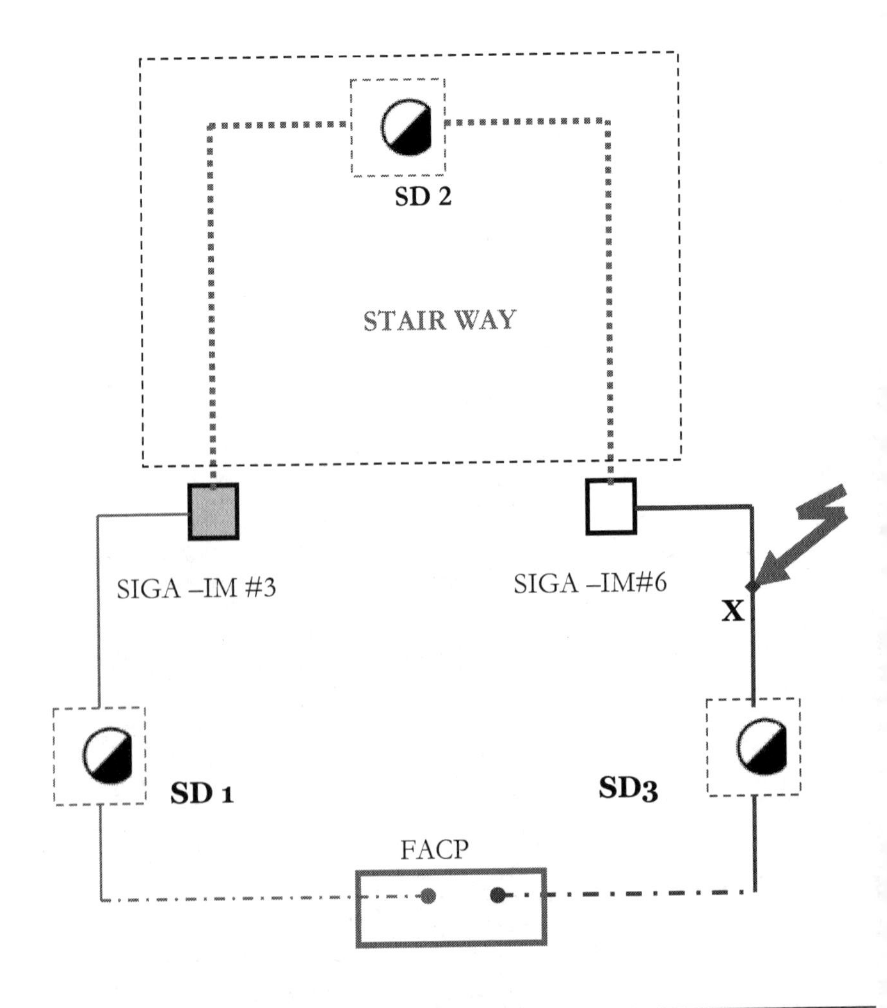

This loop is affected by the X location short but the critical area (stairs) is available with the smoke detector. More than 50 % of the loop is active so the devices located at this 50% are also active.

If the fault is removed then the module will be able to repair the entire loop, allowing the system to work according to its design.

8. IM#3 –CLOSE
9. IM#6- CLOSED
10. DS #1;2;3 – WORKING
11. The initiation loop – ACTIVE

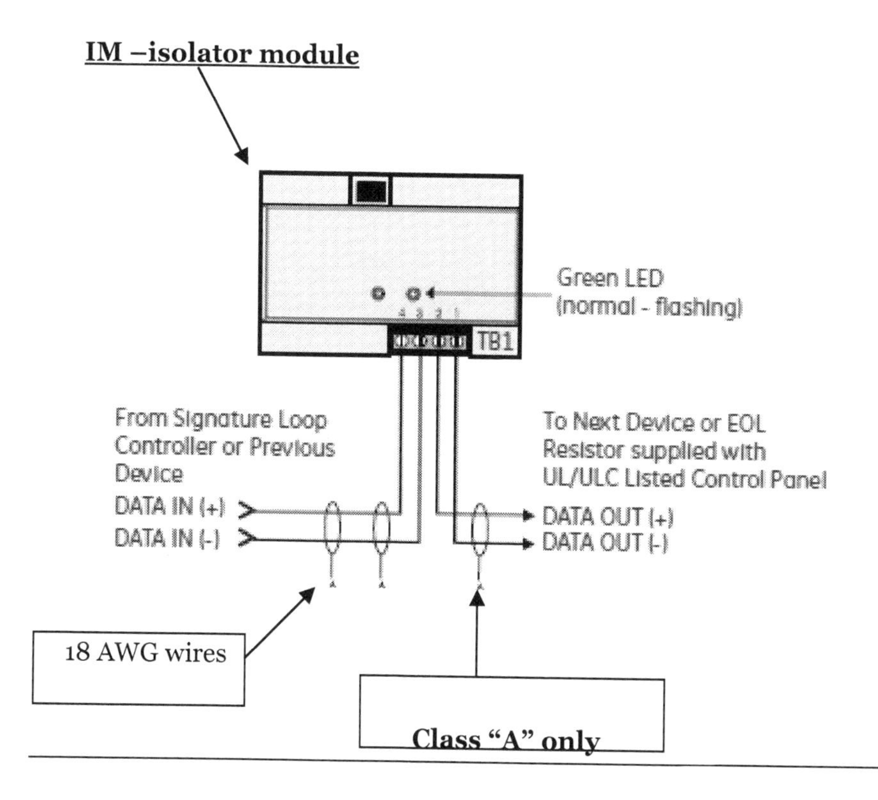

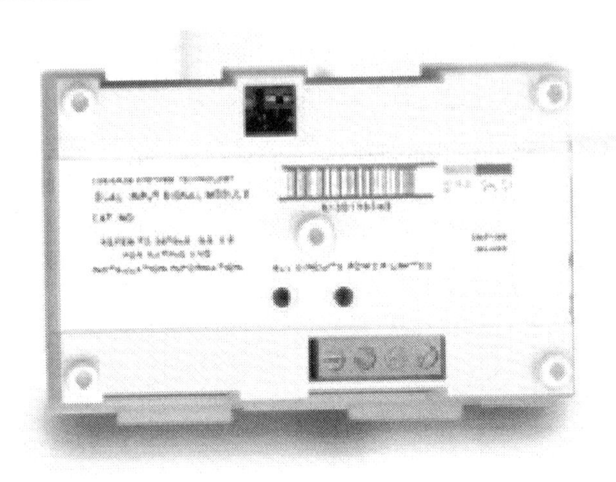

"LIVE "Isolator Module (IM)

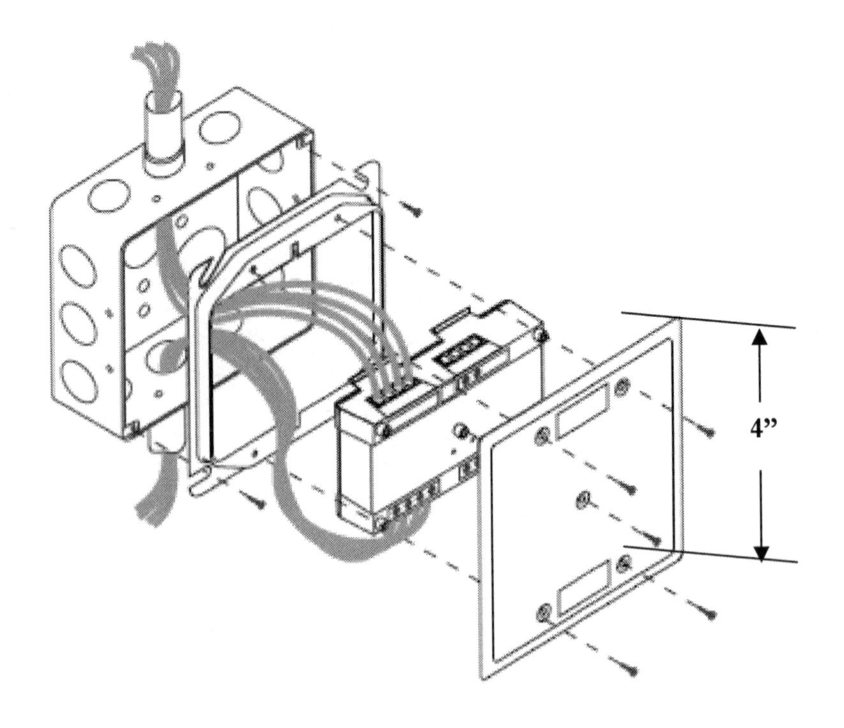

This isolator module SIGA –IM will fit into a **4" square metallic box.** The box should be at least 2 ½" deep. The device is provided with a suitable cover to fit on the 4 "x 4" box. The terminal block will permit termination of the #12 AWG and # 18 AWG wires.

Your notes:

..........................................................................................

..........................................................................................

..........................................................................................

..........................................................................................

..........................................................................................

..........................................................................................

------------------------------------------------------------------------

------------------------------------------------------------------------

------------------------------------------------------------------------

------------------------------------------------------------------------

------------------------------------------------------------------------

------------------------------------------------------------------------

------------------------------------------------------------------------

------------------------------------------------------------------------

------------------------------------------------------------------------

------------------------------------------------------------------------

------------------------------------------------------------------------

------------------------------------------------------------------------

## Elevator Homing Device

Elevators are included in most buildings with more than one floor.

The elevator and its related equipment will operate according to special procedures when a fire occurs. An elevator cannot be permitted to stay in use or to have people locked in a smoke-filled area. For these reasons there are things you need to consider when you install or maintain a fire alarm system. The elevator homing device is part of the elevator's equipment and provides information to the elevator controller in order to bring the elevator's cabin to a designated floor during a fire alarm or emergency event. If the information from the fire alarm initiation circuit or FACP is that the home floor of the elevator is affected, then the elevator homing device should be able to avoid that particular floor and direct the elevator to a different, pre-selected floor. As you will see, there are three important terms to know:

- Elevator homing device
- Elevator controller
- Fire Alarm System

These represent the elevator **recall system**. This is the system that manages the way elevators operate during a fire or emergency event. The elevator recall system requires at least three initiation device circuits (IDC). These are:

- The first circuit includes the smoke detectors that will initiate the elevator recall to the home floor. These detectors are located in the elevator entrance for those floors other than the home floor. They will detect the smoke on the other floors, locate the elevator and provide this information to the controller. The controller will then recall the elevator to the home floor.
- The second circuit includes the smoke detector(s) located at the elevator entrance on the home floor. If this detector detects the smoke, it will deliver a signal through the FACP to the elevator controller regarding this information. The controller will then recall the elevator to an alternate floor that is safe and without smoke.
- The 3rd circuit includes the smoke detector located in the machine elevator room and hoist way and will recall the elevator if smoke or excessive heat is detected in the elevator shaft or the elevator is no longer safe.

For increased safety there are codes requiring that smoke detectors located on these circuits be relay base detectors with normally–open contact (NO). See the diagram below:

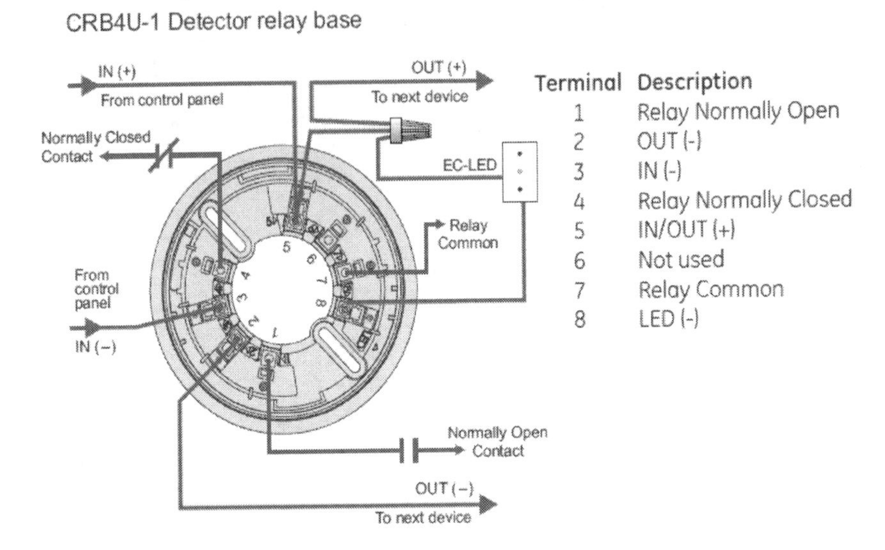


| Terminal | Description |
|---|---|
| 1 | Relay Normally Open |
| 2 | OUT (-) |
| 3 | IN (-) |
| 4 | Relay Normally Closed |
| 5 | IN/OUT (+) |
| 6 | Not used |
| 7 | Relay Common |
| 8 | LED (-) |

This NO contact will "inform" the controller located in the elevator machine room. The controller will generate the recall elevator task. All these detectors can be integrated into the same loop when the initiation devices are intelligent, addressable devices. Control Relay (like SIGA-CR) can be installed in case you don't use a relay base detector. The wiring diagram for the control relay indicates the loop terminations (IN/OUT) and the relay terminations. (NO; NC; COM)

## Control Relay

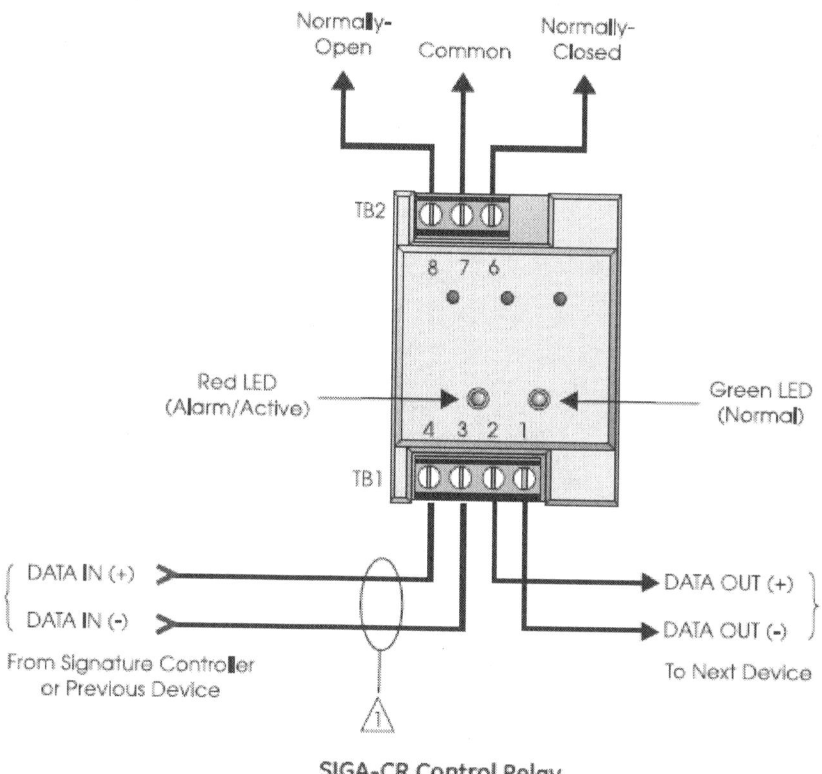


SIGA-CR Control Relay

This device should be installed in an accessible location where it can be easily reached for maintenance. This is the detail you should use when required to consider the installation of an addressable module. Obviously you'll install a support for this box (Single-gang electrical box, 2.5 "deep) unless there is a surface mounting detail on the wall or an existing support. Please also observe the conduit's connector and the spare length of wire inside the box. These are items you need to consider when delivering this product to ensure you provide easy maintenance, high quality work and an all-around functional and reliable system. The faceplate of this device is designed in such a manner that the green LED & red LED will be visible when flashing. Green flashing indicates a normal system swhile red indicates an "alarm" signal.

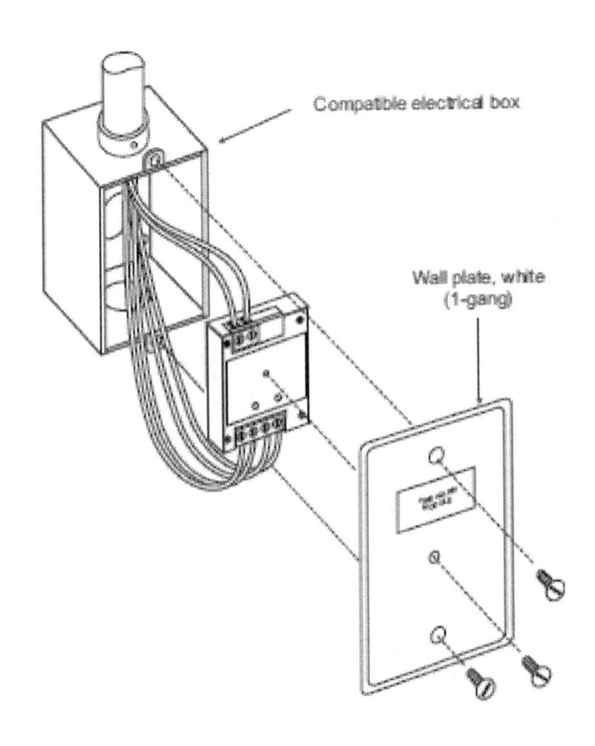


**Magnetic Lock**

In previous chapters when we had a Fire Alarm System description we mentioned that between the system's components there are AUXILIARY COMPONENTS of the system: *Also there are components that are auxiliary to the system, they are not supervised by the FACP but they are controlled by the FACP. Such auxiliary devices are:*

- *Door holders release (Electro-Magnets on the wall to hold the door open will release the door so that it closes, creating a barrier between the fire and the area)*
- *Door magnetic lock release and others, such as:*
- *Elevator recall*
- *Pressurization fans*
- *Fan shut down devices*

Some rooms in different types of buildings will require controlled access due to security reasons. Some of these doors will have magnetic locks to control the access through the door. These magnetic locks will release the door when needed under the normal operating conditions of day to day business. A FIRE ALARM EVENT is not a normal operating condition in any building and for this reason the fire alarm system should unlock all doors during such an event so that people can be evacuated. The next picture shows the "interconnection "between the fire alarm and security /access systems.

Interconnection box with other systems

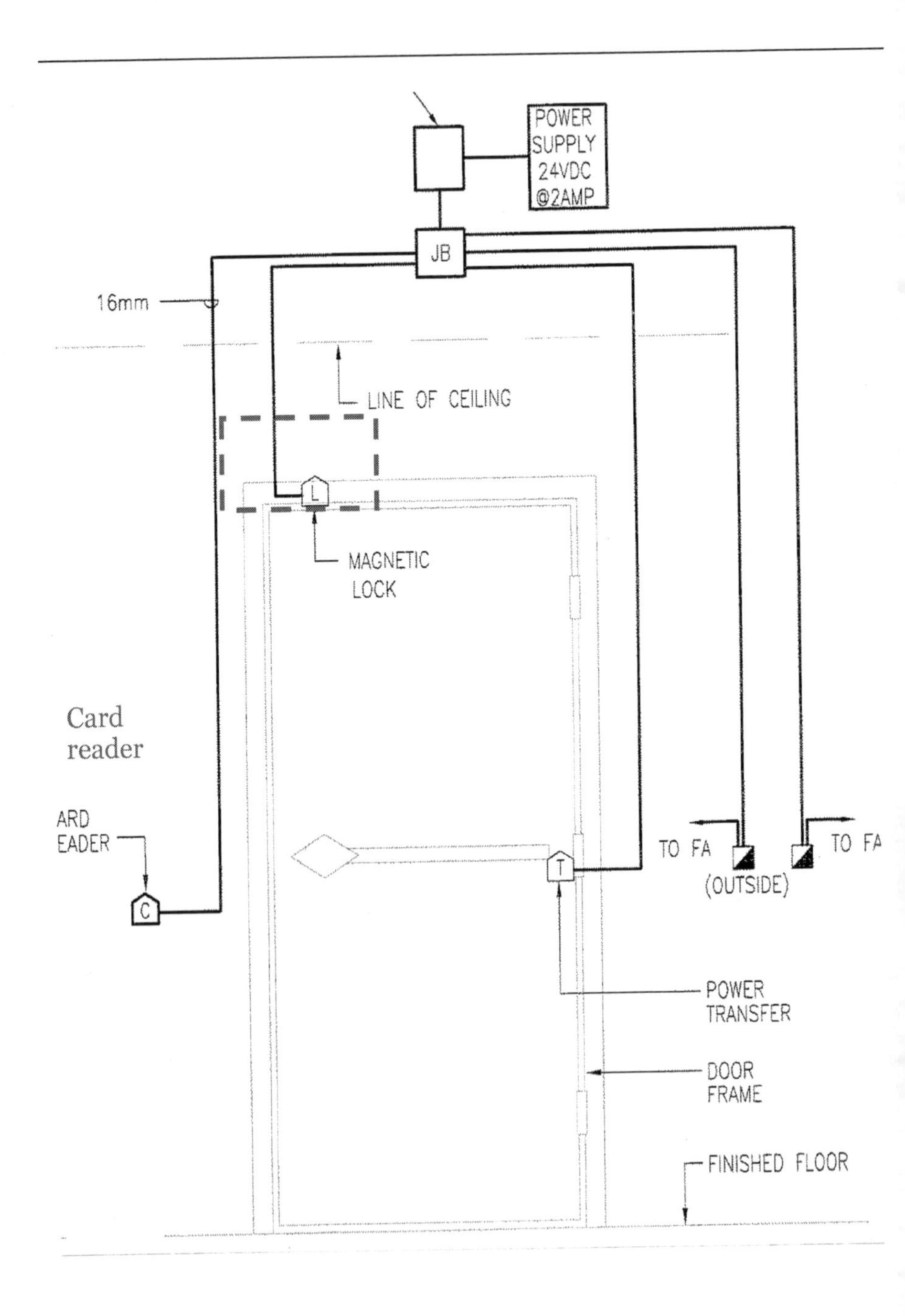
POWER
SUPPLY
24VDC
@2AMP
JB
16mm
LINE OF CEILING
L
MAGNETIC
LOCK
Card
reader
ARD
EADER
C
T
TO FA
TO FA
(OUTSIDE)
POWER
TRANSFER
DOOR
FRAME
FINISHED FLOOR

These are the parts of the door's magnetic lock.

## Door Holder

Some doors should automatically close to provide safe separations from the fire between areas. In order for this to happen, the magnets on the wall holding the door open (the door holders) will release the door. Once released, the door will mechanically close, ensuring that there is a fire separation in the area. These devices look like those pictured below:

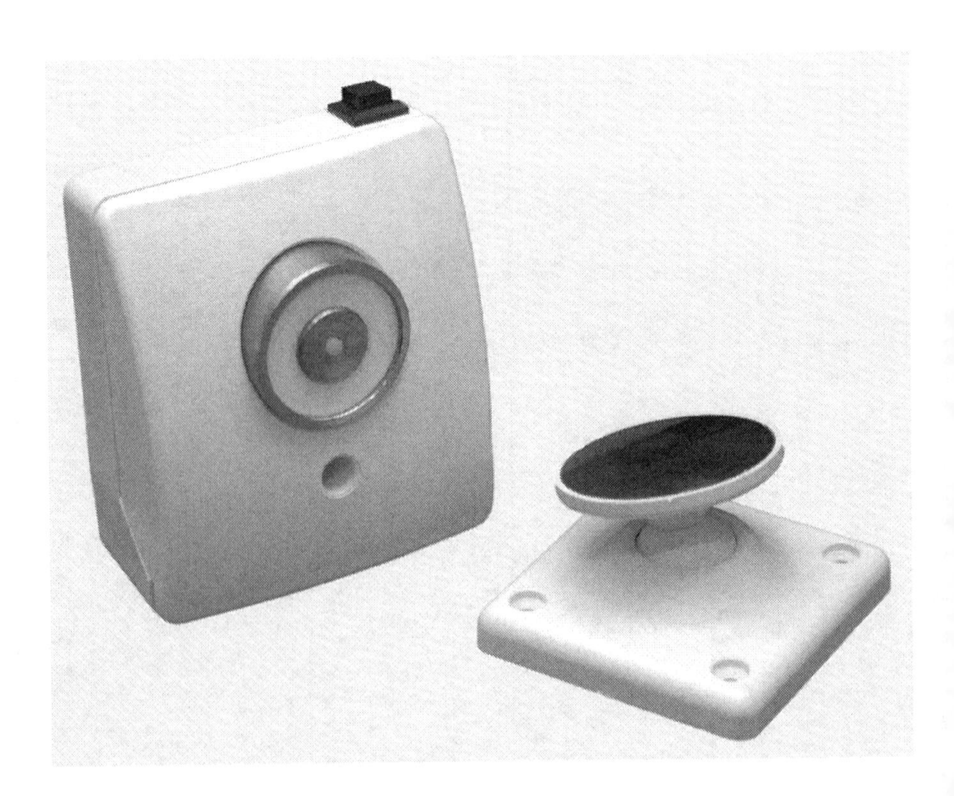

http://www.security.honeywell.com/star/hsce_files/Files/O-DoorHolder.htm

Read all you need to understand the door holder purpose!!!

## TOP VIEW DOOR PROFILE

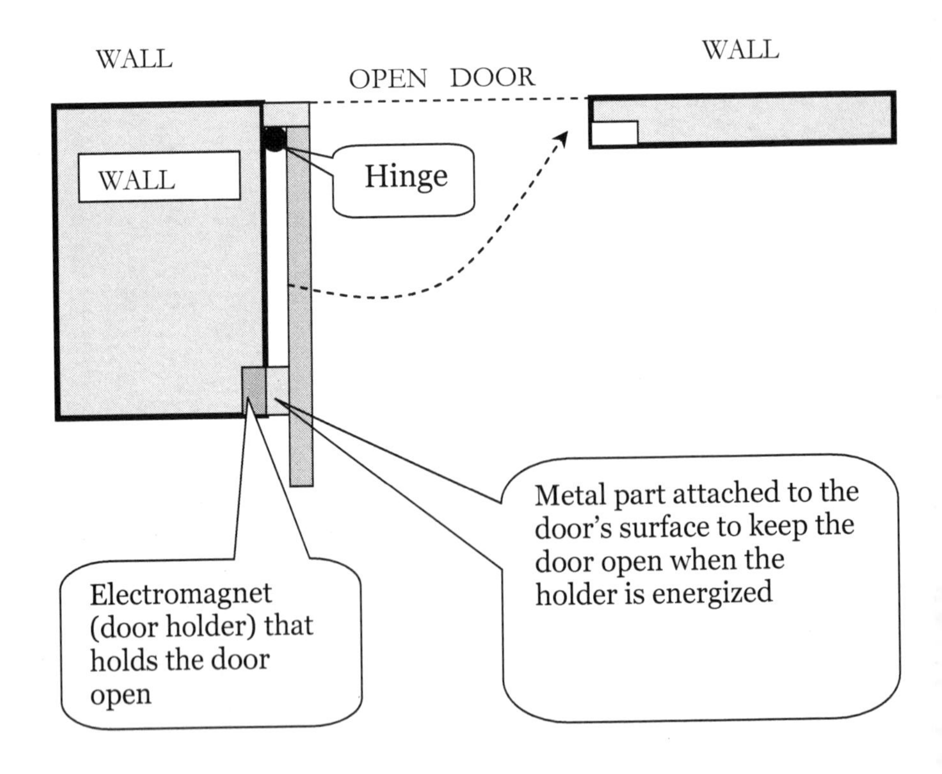


How this operates:

When a fire hazard occurs and the alarm signal is present, the door holder will lose power (usually 24 volt DC).

At that moment the door holder will demagnetize, causing the door to close under the pressure of the tensioned spring at the hinge. The door thus creates a fire separation for a certain period of time (if the design has requested this as a way of providing a safe area to reduce the possibility of a fire spreading). For a clearer understanding of the situation, I will describe the situation after a fire event:

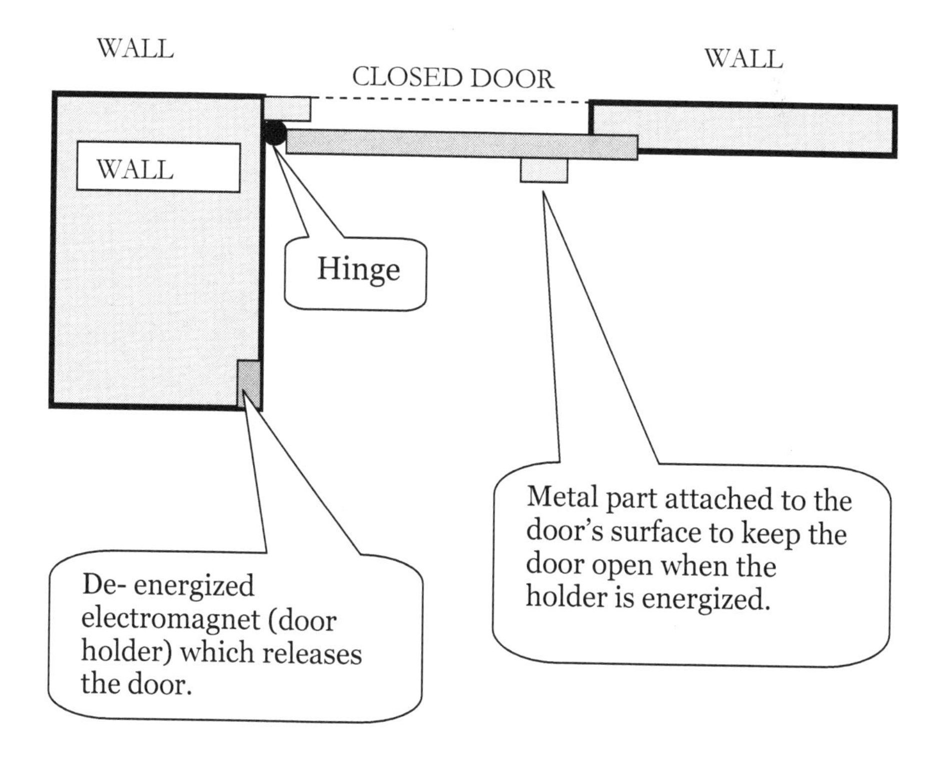

# Door holder is connected (ON)

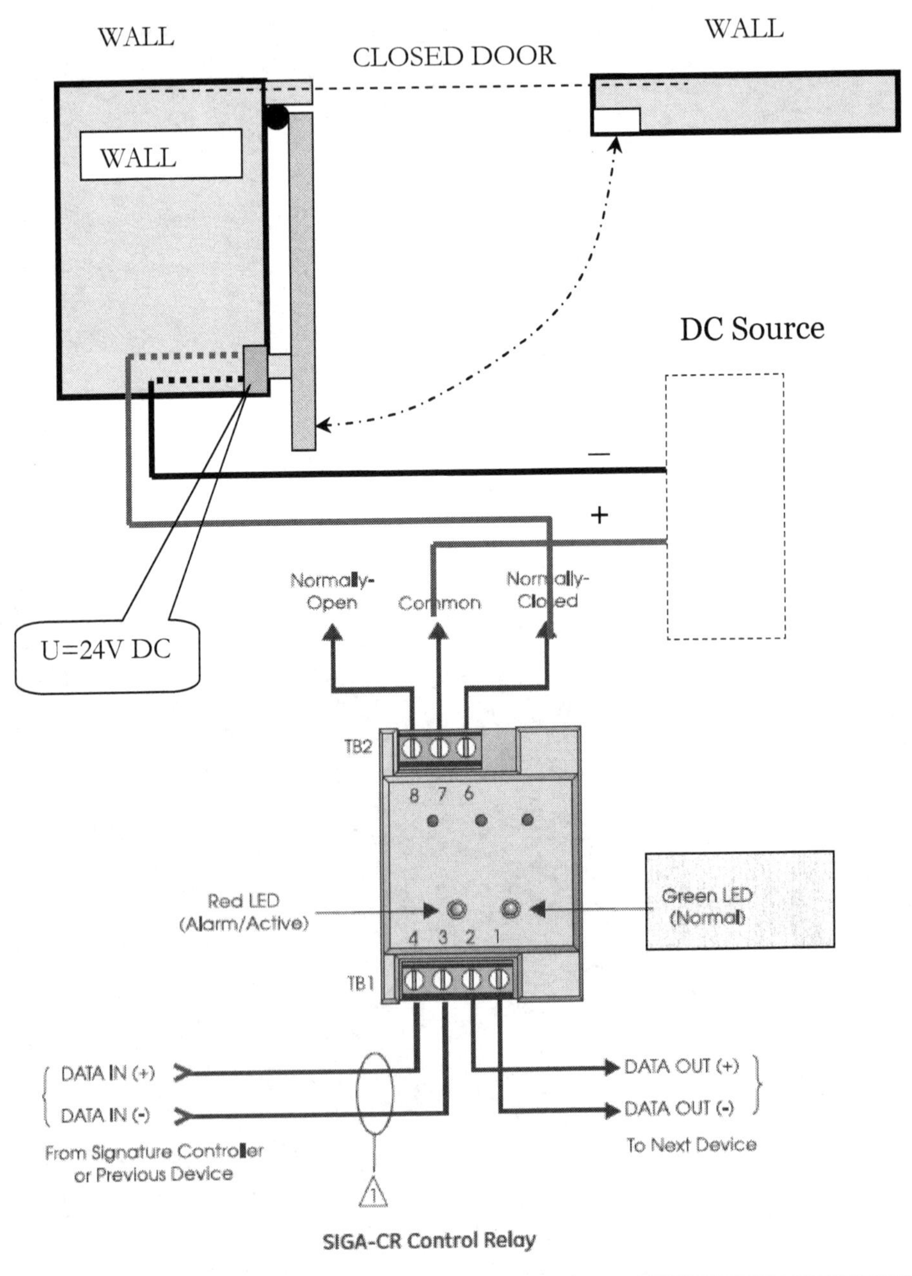

# Door holder is connected (OFF)

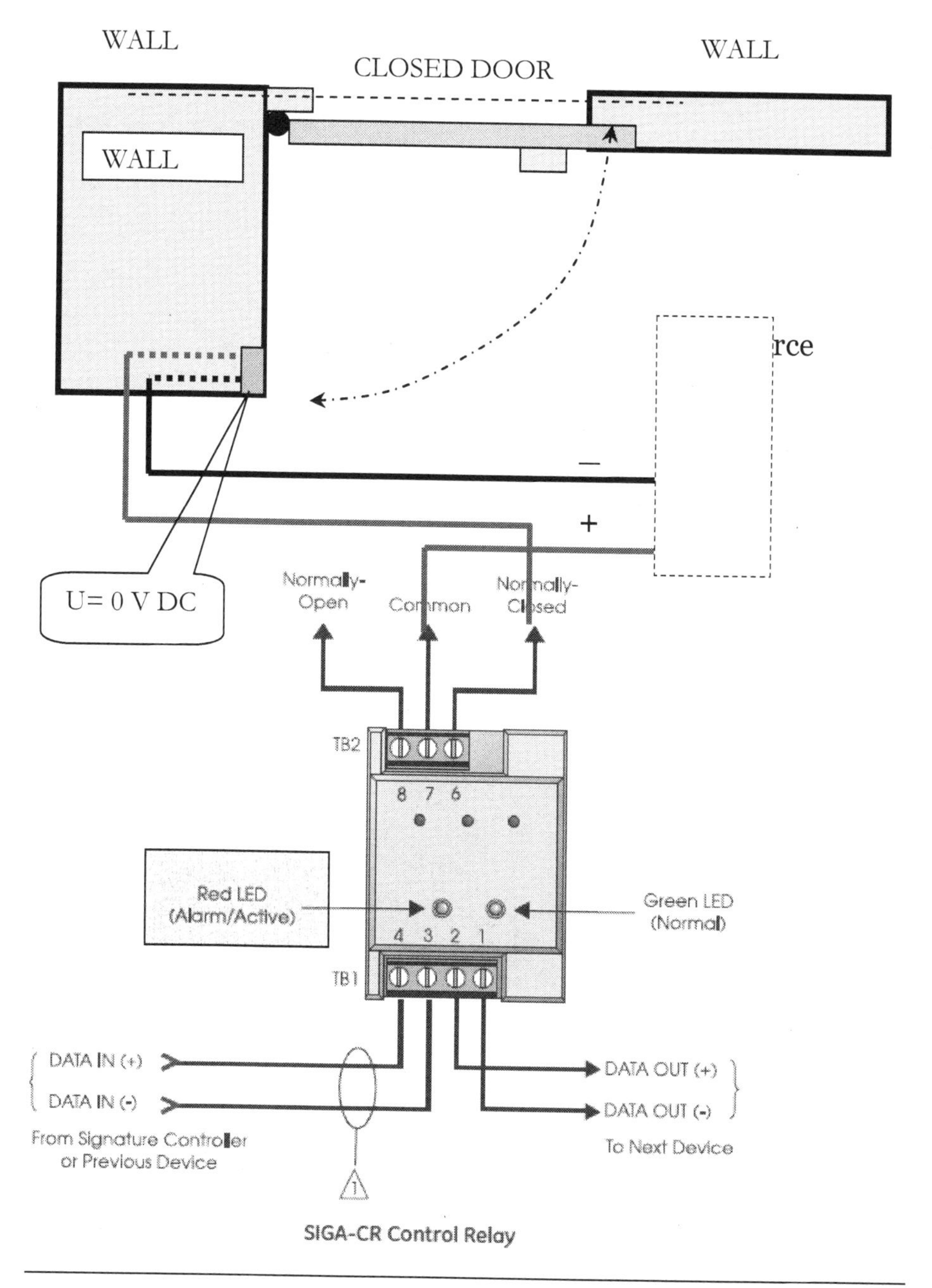

## Emergency Power Supply and Fire Alarm System

For example, imagine we are talking about a hospital. There are systems in which the power is shut-off during a fire but in other cases the generator should be able to automatically start and supply the vital loads.

In most buildings the generator is monitored and automatically starts.

The status of the generator is indicated also by the FAS

- generator running or
- Low fuel level or
- The generator's circuit is in "trouble"

The information mentioned above:

**"Running"-"fuel level"-"schematics trouble ",** will need to be monitored in order to provide dates about the availability and integrity of the generator in these areas where is required. It is supposed keep the system "under control" so it can "react" when needed. A non-responsive or "faulty" system is a major problem. Therefore, maintenance is crucial. The next few pages will demonstrate how the Generator Room is interconnected with the FAS.

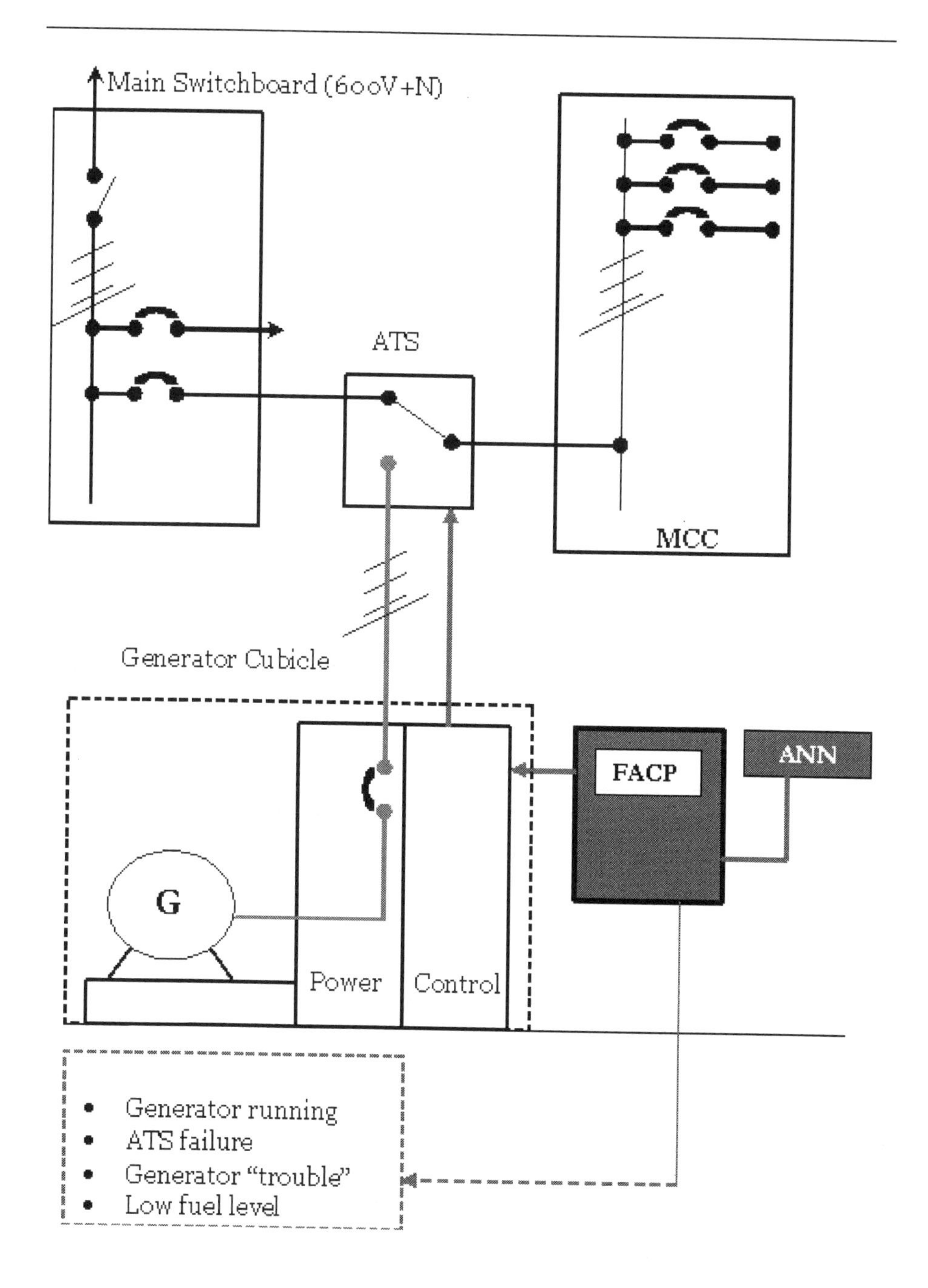


ATS- stands for: Automatic transfer switch

## Devices on the Notification Appliance Circuit (NAC) for Addressable FAS.

The Notification and Appliance Circuit (NAC) includes devices that send out notifications about a fire alarm event (horns; strobes; combination horn/strobe; speakers). *The notification and appliance circuit is limited by the number of devices on it. The way these special devices work is indicated on the diagram below which represents a class "B" typical NOTIFICATION (NAC ), 2-wire circuit (This is not a loop because the loop should go back to the fire alarm control panel or FACP). The radial lines (which spread from the panel but do not come back to the panel) are the Notification & Appliance Circuit (NAC). These lines include devices that send out notifications about a fire alarm event (horns; strobes; combination horn/strobe; speakers). The NAC (devices like horn/strobes/speakers) will notify occupants of the need to evacuate the area should an emergency/fire event occur on the system.

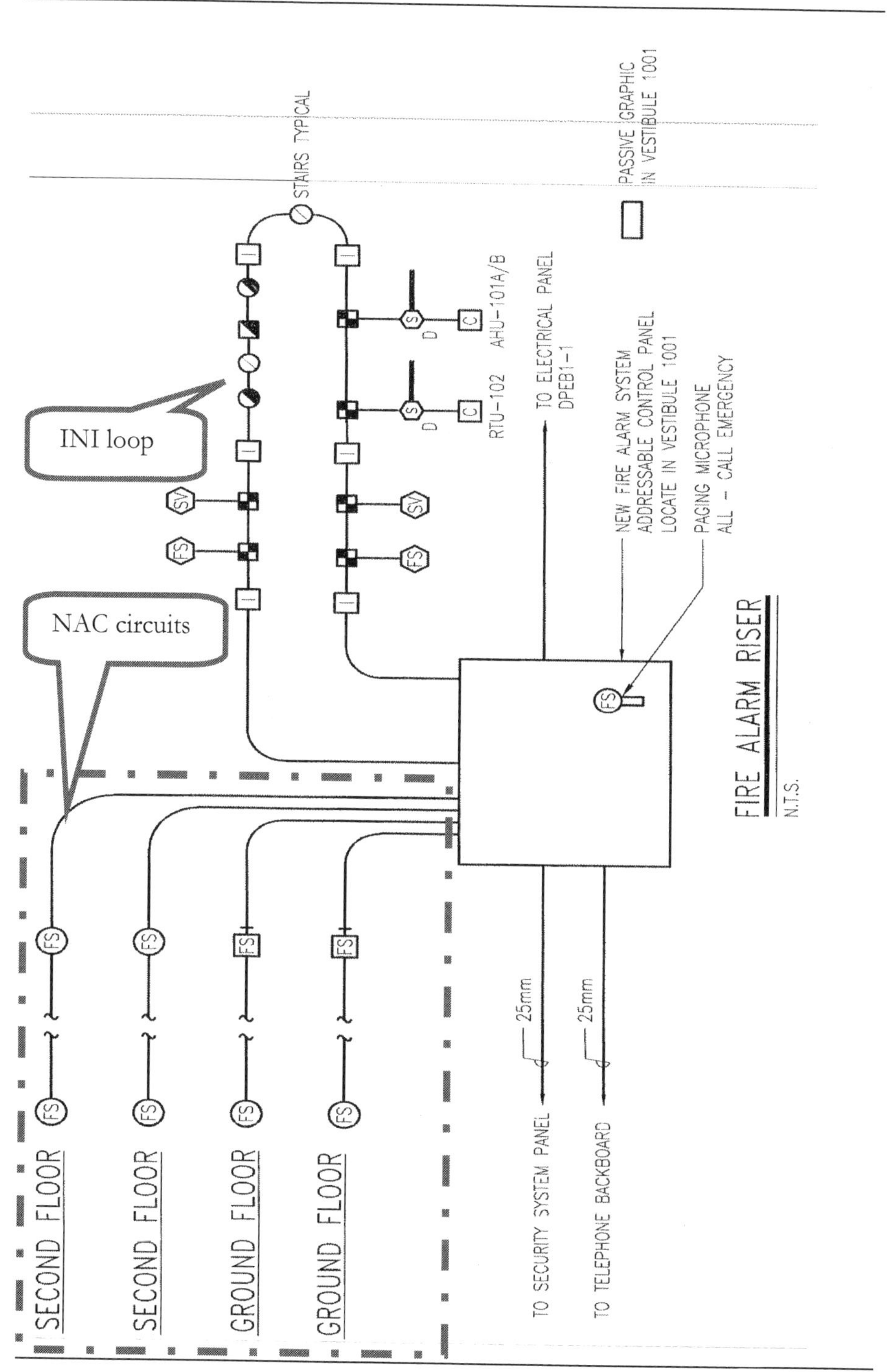
INI loop
NAC circuits
STAIRS TYPICAL
PASSIVE GRAPHIC
IN VESTIBULE 1001
RTU-102
AHU-101A/B
TO ELECTRICAL PANEL
DPEB1-1
NEW FIRE ALARM SYSTEM
ADDRESSABLE CONTROL PANEL
LOCATE IN VESTIBULE 1001
PAGING MICROPHONE
ALL - CALL EMERGENCY
FIRE ALARM RISER
N.T.S.
SECOND FLOOR
SECOND FLOOR
GROUND FLOOR
GROUND FLOOR
25mm
25mm
TO SECURITY SYSTEM PANEL
TO TELEPHONE BACKBOARD

## Fire Alarm Horn

- Positions three (3) and four (4) are indicating elements such as the FIRE ALARM HORN or STROBE or a combination HORN & STROBE.

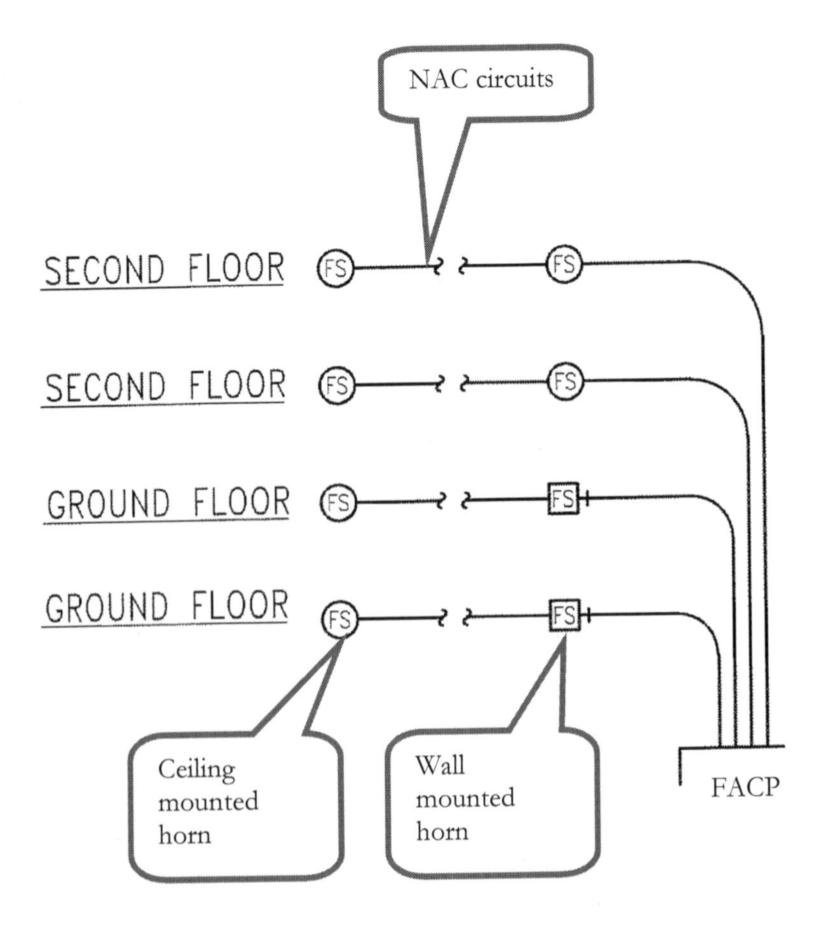


The devices indicated in the diagram are ceiling or wall mounted horns or speakers or a combination horn/strobe (acoustical and visual) signal. These circuits will be terminated with the EOL (end-of line resistor). The role of the EOL resistor is to monitor the circuit. This small resistor will confirm the circuit's status. In previous chapters the resistor was described in further detail. If you feel the need to refresh your memory, go back and read this information.

For those of you who've never had the chance to see such devices, please have a look at the photos below. They will likely look familiar as they are present in schools, malls and all public buildings. These devices are designed to save lives and protect facilities from disasters.

Horn & strobe

Speakers

## Fire alarm horn & strobe

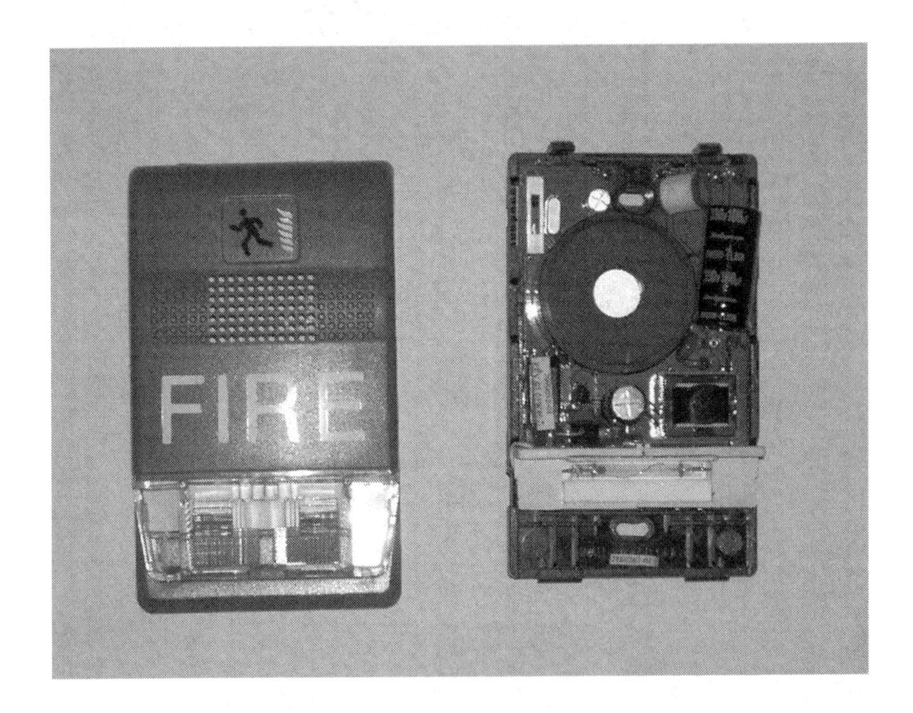


The photo above shows the GE Security Horn & Strobe. The cover is removed to make the inner workings of the device visible. The horn is an audible Notification appliance and a high quality device. This device may be set to sound at a temporal or steady tone. The tone level can be either 98 or 102 dBA (dB stands for decibel and is the common unit used to measure noise levels or magnitudes and A is the symbol used to denote the specific response a human ear would make to a sound). The strobe is the visible Notification appliance that is self-synchronized to flash at 1 FPJ (flash per second) when voltage is applied. The light intensity is measured in candela (cd) and can operate within a range of: 15/75; 30; 60 or 110 cd. The next diagram will show the wiring connection when the horn and strobes are on the same circuit.

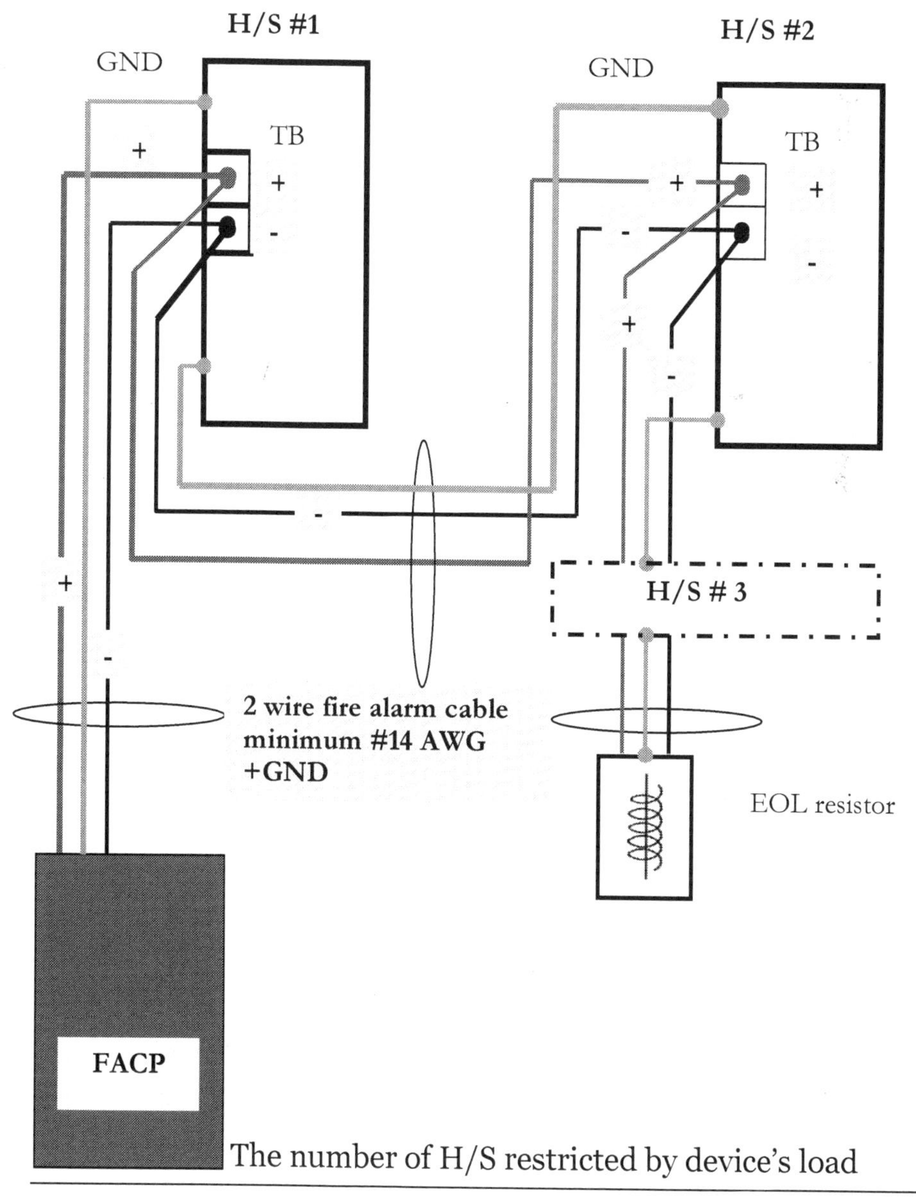


The number of H/S restricted by device's load

TB – terminal block, H/S – combination horn & strobe

GND –bonding wire to grounding system through FACP. There are situations where the horn and strobe is connected on different circuits. In this case, a 4-wire cable should be used-two wires for the horn and two wires for the strobe. The wiring diagram can be found on the next page.

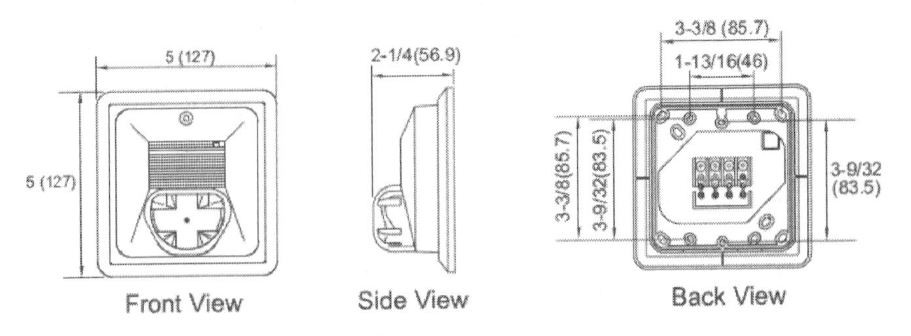


MIRCOM Horn/Strobe –FHS 240(TB with 4 terminals)

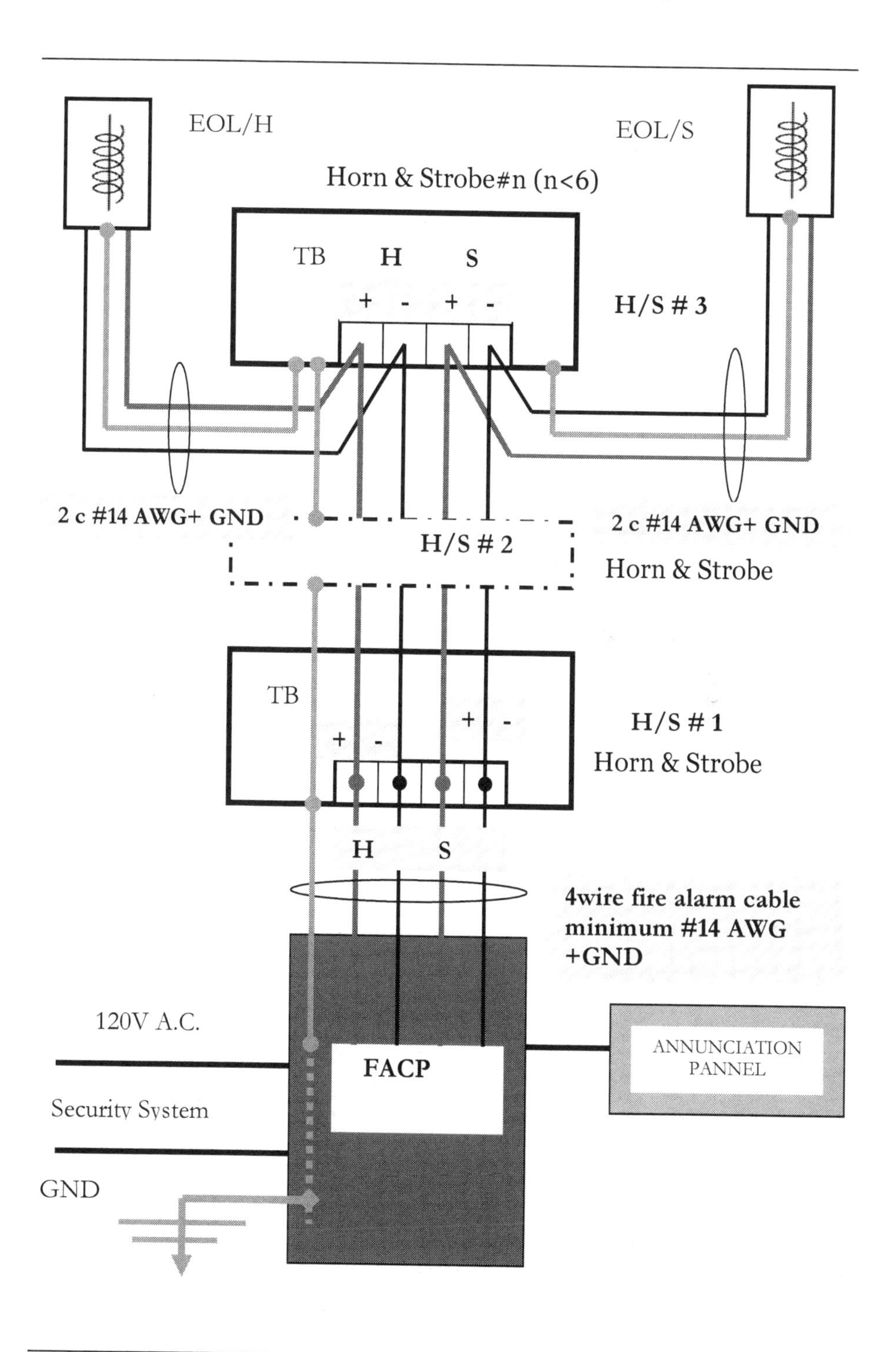
EOL/H
EOL/S
Horn & Strobe#n (n<6)
TB
H
S
+
-
+
-
H/S # 3
2 c #14 AWG+ GND
H/S # 2
2 c #14 AWG+ GND
Horn & Strobe
TB
+
-
+
-
H/S # 1
Horn & Strobe
H
S
4wire fire alarm cable
minimum #14 AWG
+GND
120V A.C.
FACP
ANNUNCIATION
PANNEL
Security System
GND

## Fire alarm strobe

4MS Series

This device is a strobe whose manufacturer is GE Security. The strobe is available in 15/75 and 110 candela models and is designed to alert people in the event of a fire. It works in the range of 20 to 31 volt dc and must be connected to a signal circuit with a constant voltage. The device has a diode that allows for the supervision of the circuit. This device can be installed on ceiling and walls. Indoor and outdoor installation is also possible but the proper box needs to be installed. Outdoor installation will require a metal weatherproof box to protect this

electronic device whereas with indoor installation, only a regular 4x4 box, as indicated here, is needed.
The photo shows a 4 x 4 metal box. Some boxes are provided with a grounding pigtail as you can see in the picture below.

There is an intermediate plate that functions much like an adapter to make the device fit into the box. The same detail I used for the IM (isolator module) is applicable here as well. Ignore the control module and the faceplate in order to better understand how this device will fit into the box.

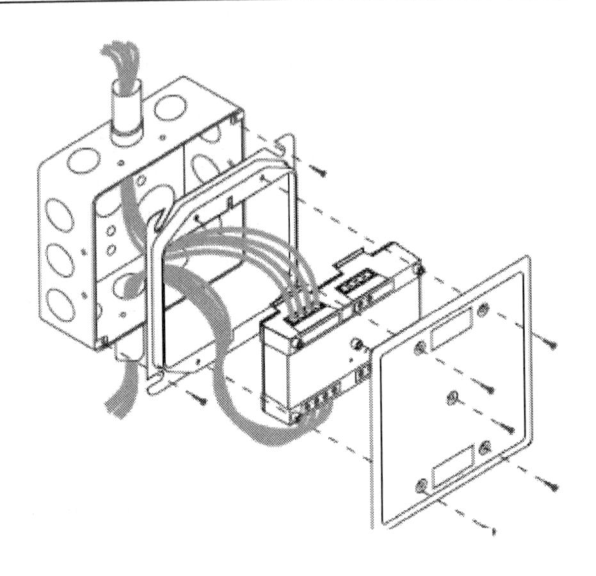

The connectors for conduit look like:

FA= fire alarm (connectors type)

Wiring Diagram for a fire alarm strobe:

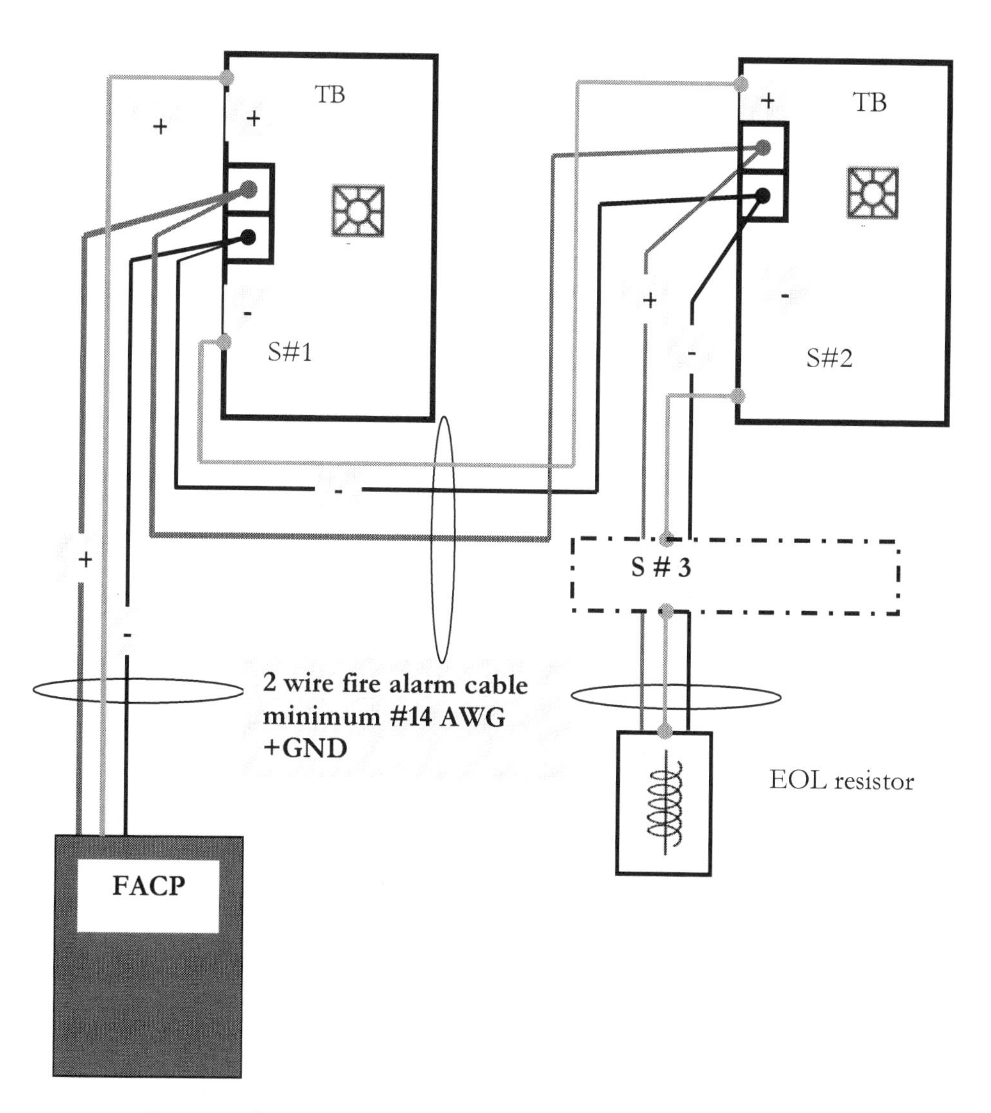


Note: The number of H/S restricted by device's load

TB – terminal block; S – strobe;

GND –bonding wire to grounding system through FACP

## Visual and Audible Signal Synchronizing

In fire alarm systems, the synchronization of the visual and audible signals released by the devices installed on the Notification and Appliance Circuit (NAC) is likely. In other words, it is important that the light is seen at the same time the alarm is heard. The visual signal (the strobe light) from the strobe should also be synchronized. When the system's installation is complete, it then needs to be tested in order to verify that the installation meets the design requirements. The synchronization of the visual and audible signals requires verification. There are electronic devices installed in the fire alarm panel that will provide synchronization. This chapter's aim is to familiarize you with this device, even if the supplier of the FACP (fire alarm control panel) includes this in the furniture of the panel.

I was working with Edwards's fire alarm control panels and remember using Genesis Signal Master: G1M-RM. What this device is doing is converting one audible NAC and one visual NAC (notification circuit) into a one NAC 2-wire circuit with Horn & Strobe.

The G1M-RM is providing the synchronization and inaudible "silence" function. I recommend you to study this device on www.gesecurity.com in order to further understand it. There are many applications and wiring connections for this device.

The bottom line I need you to keep in mind is this: for synchronization to occur, on the NAC circuit you'll need to use a "signal master" that is able to provide the synchronization of devices on the circuit as well as supplying the Horn & Strobe on 2-wire circuit!

The advantages of this are:

1. Synchronization of strobe and independent control of horns.
2. Features on the 2-wire circuit

I will include a diagram in which this device is installed close to FACP to synchronize the horns and to provide the silence function for horns. Again, the horn and strobe is the same unit in this diagram. There are provisions included by the manufacturer to allow the installation of the Signal Master into the same box as the first device (horn/strobe)

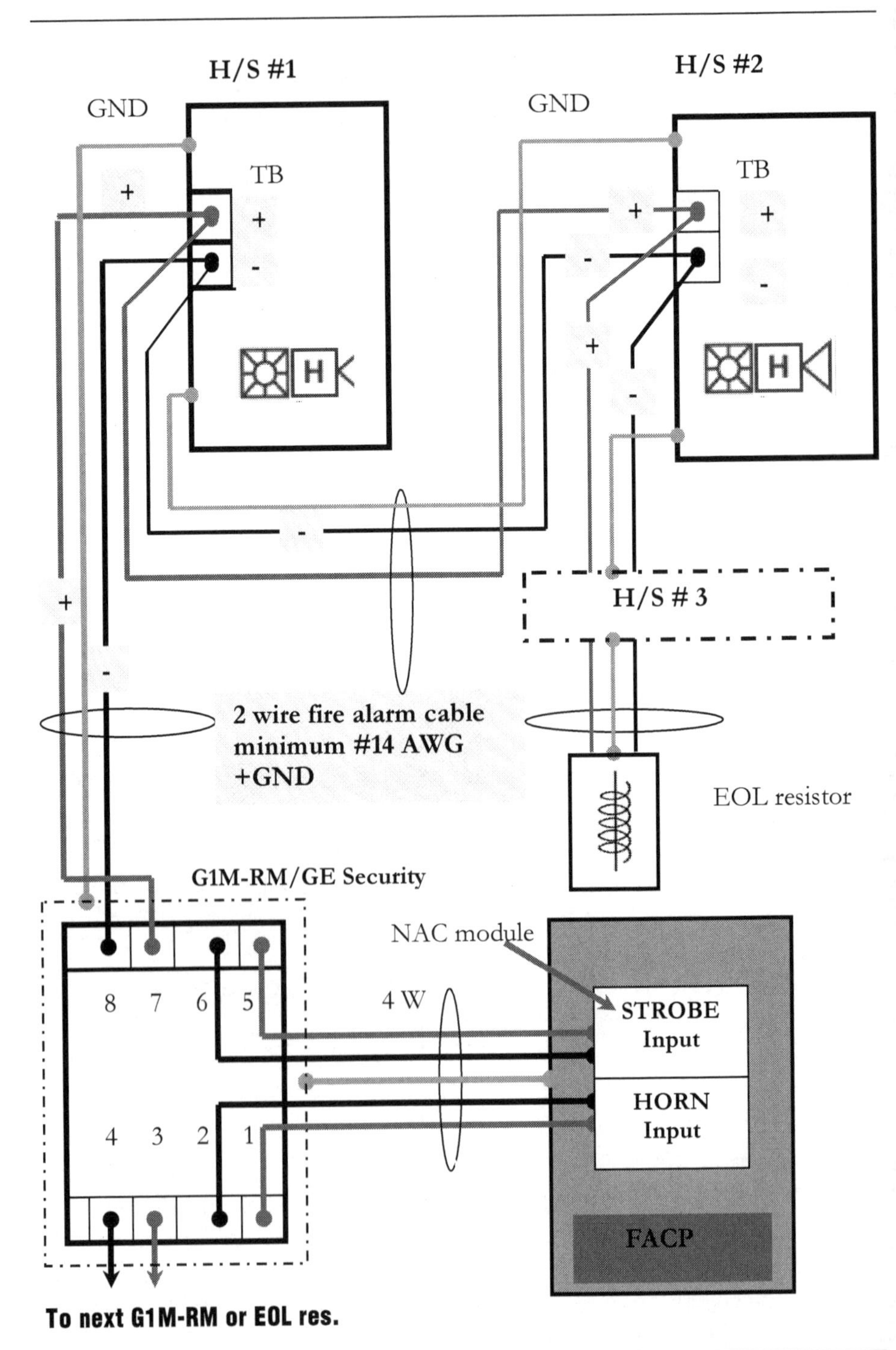
H/S #1
H/S #2
GND
GND
TB
TB
+
+
+
+
-
-
-
-
+
-
H
H
H/S # 3
2 wire fire alarm cable
minimum #14 AWG
+GND
EOL resistor
G1M-RM/GE Security
NAC module
8 7 6 5
4 3 2 1
4 W
STROBE
Input
HORN
Input
FACP
To next G1M-RM or EOL res.

If there is a requirement for the NAC circuits to be in class "A" wiring distribution that is not a big deal. Just bring the wiring that ends at the EOLR to the Fire Alarm Control Panel (FACP). The resistor is going to be installed on the terminal block. I've had some projects where the designer's requirements were to bring all the EOLR into one place beside the FACP and to label them accordingly.

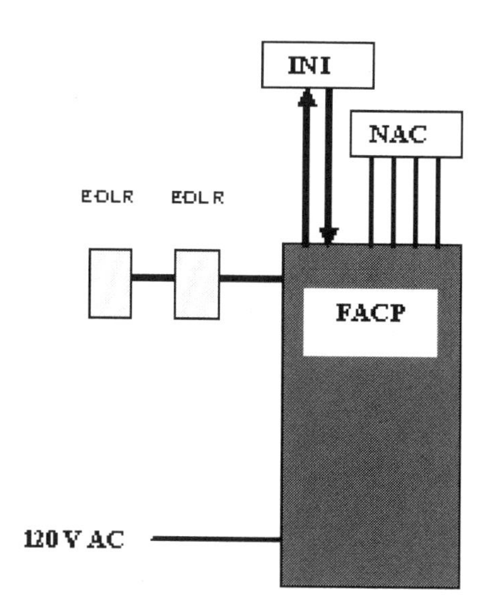


The INI loop represents one conduit protecting the wires and going from the FACP to the first device and one conduit from the last device coming to the FACP.

## IM (isolator module) for Notification Appliance Circuit (NAC)

Until this point information has been provided about different products I've installed during projects. I've also attempted to introduce you to some of the items available on the market whilst also trying to make the installation process and roles of different devices comprehensible. For example, one project I worked on was a retirement home. The fire alarm system was a very comprehensive one and, at that time, I used MIRCOM products. You may "Google" them on the net to find out more. I will refer to the devices to be installed on the NAC circuit since I found them very interesting. I want to share this experience with you. On the NAC circuit there is a similar device, we were discussing in previous chapters at the INITIATION circuit (the Isolator Module: SIGA-IM). You'll remember that this device was able to isolate the portion of an initiation circuit that is defective whilst keeping the important parts of the loop available.

The NAC circuit also includes a device that provides high reliability for the NAC circuit by making the audible (horn) and visible (strobe) devices available. As per design in the retirement home, there was a requirement to install the smoke detectors into the rooms as part of the INI loop while the temporal horn was to be installed into the rooms as part of the NAC circuits. In the corridors there was a requirement for a combination of horn and strobe. The wiring circuit was a class "A "circuit. For each of the two adjacent rooms there was a requirement to install an Isolator Module.

To "copy" this model I included in the project two rooms on the second floor to act as a visitor's room similar to an apartment area. I am referring to rooms R 212& R214.

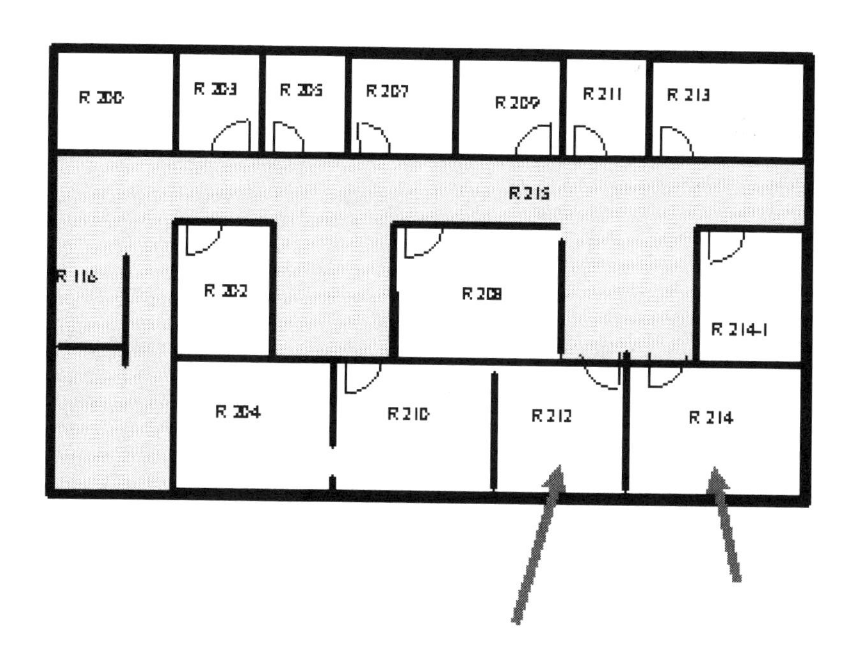

## NAC class A-Signal circuit isolator module (SCISM)

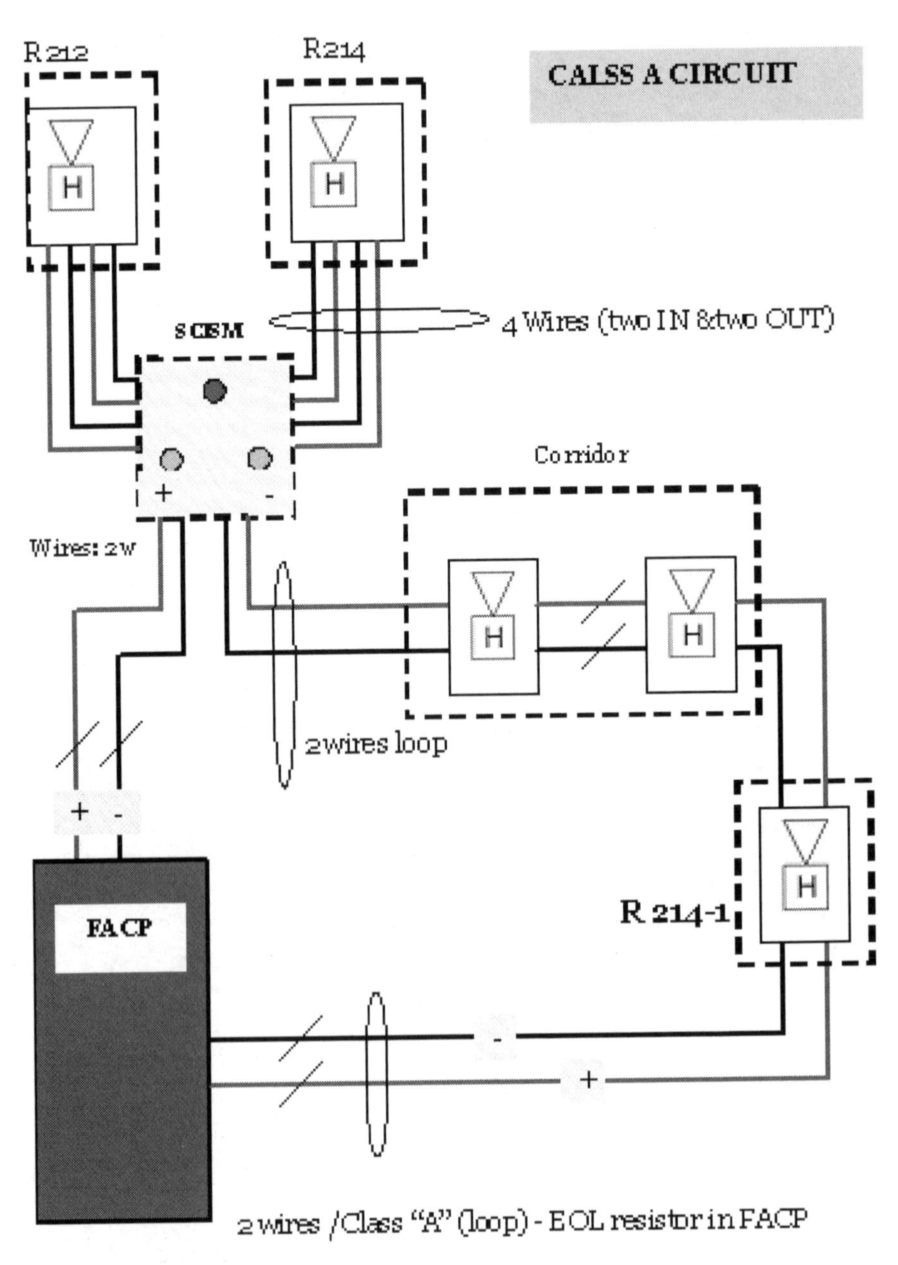

How this device works: Any damage or circuit opening of the room R212 and/or R214 **SCISM will isolate** the horns from affected circuit and will keep the others able to respond to any signal triggered by the initiation circuit.( Horn # 3;Horn #4 or horn located in room R214-1.In case the minus or positive line from horn will have trouble the amber LED will flash. This is important information for maintenance staff. IN case of alarm signal the red LED will flash.

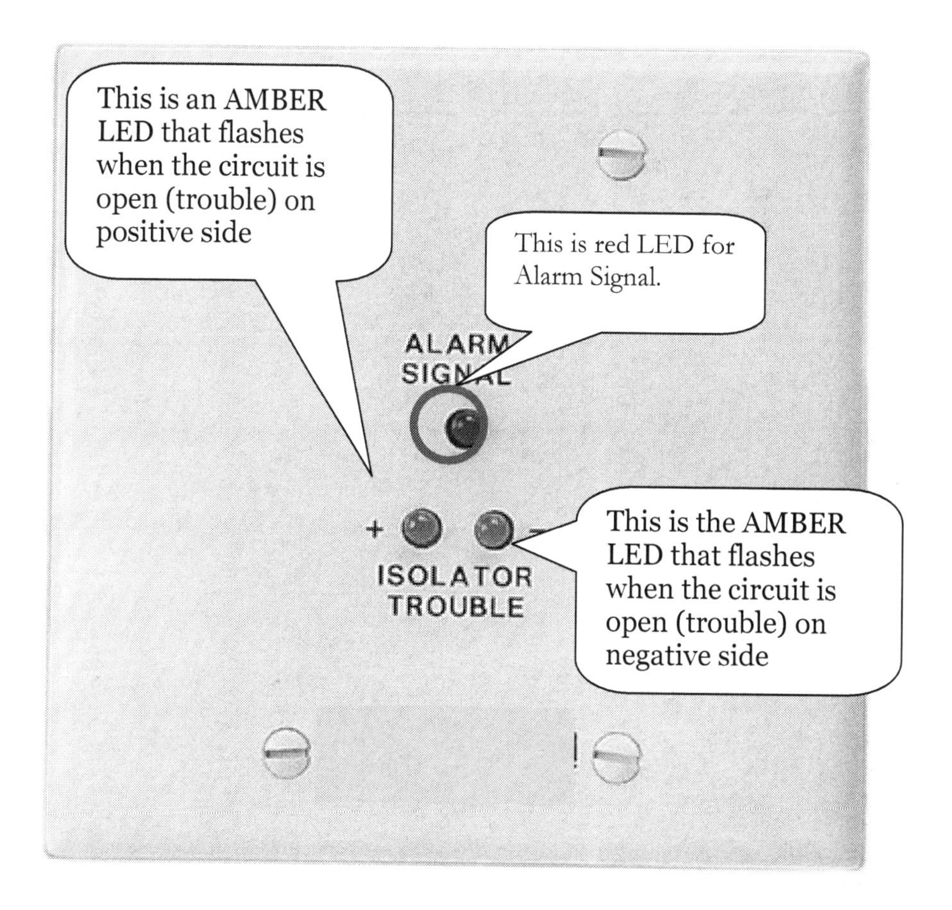

## NAC class B-Signal circuit isolator module (SCISM)

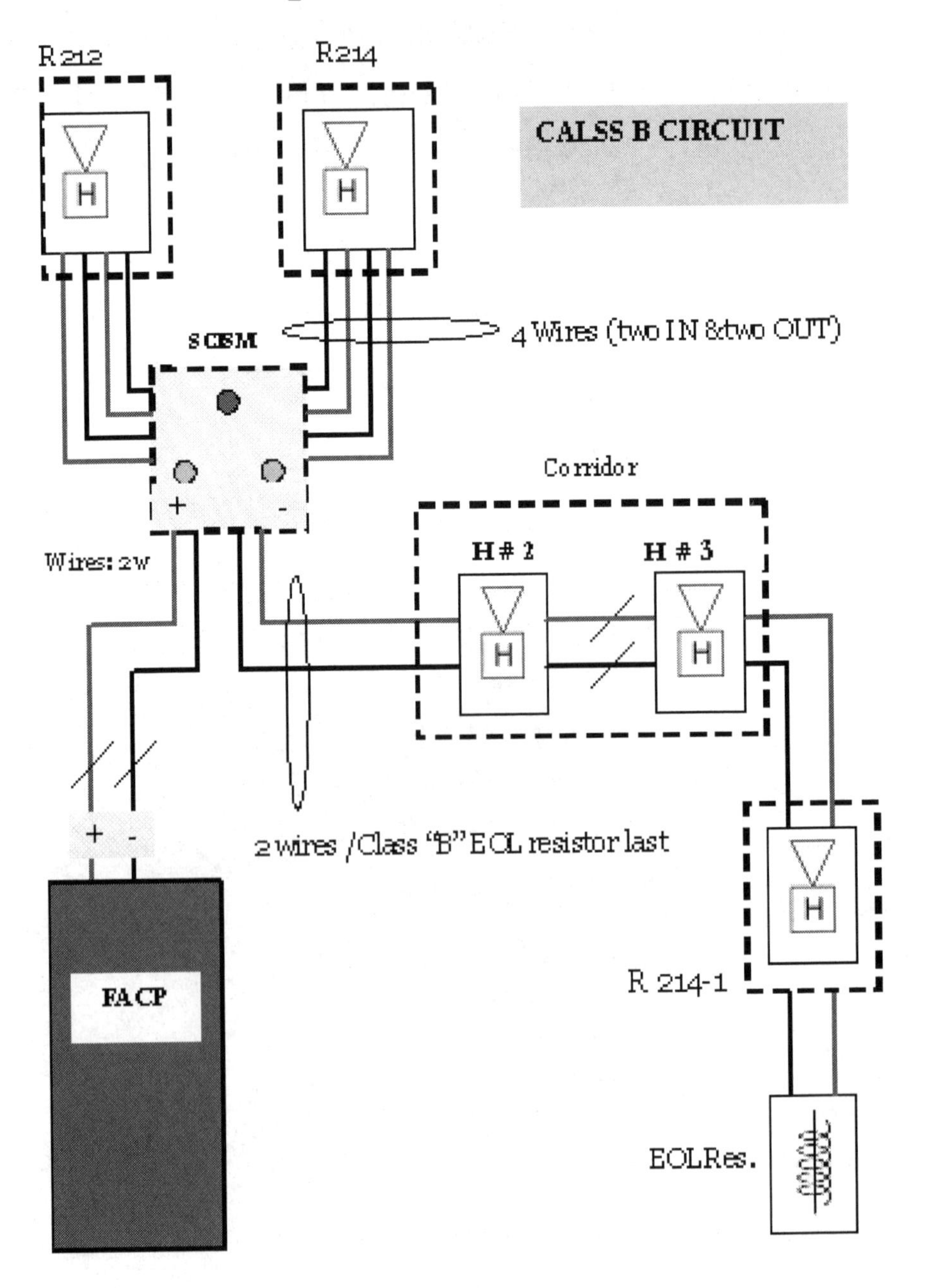

How this device works: Any damage or circuit opening in rooms R212 and/or R214 will result in the **SCISM isolatating** the horns from the affected circuit whilst keeping the others so they are able to respond to any signal triggered by the initiation circuit. (Horn # 3;Horn #4 or horn located in room R214-1). In case the minus or positive line from horn has trouble, the amber LED will flash. This is important information for maintenance staff to know. IN case there is an alarm signal, the red LED light will flash.

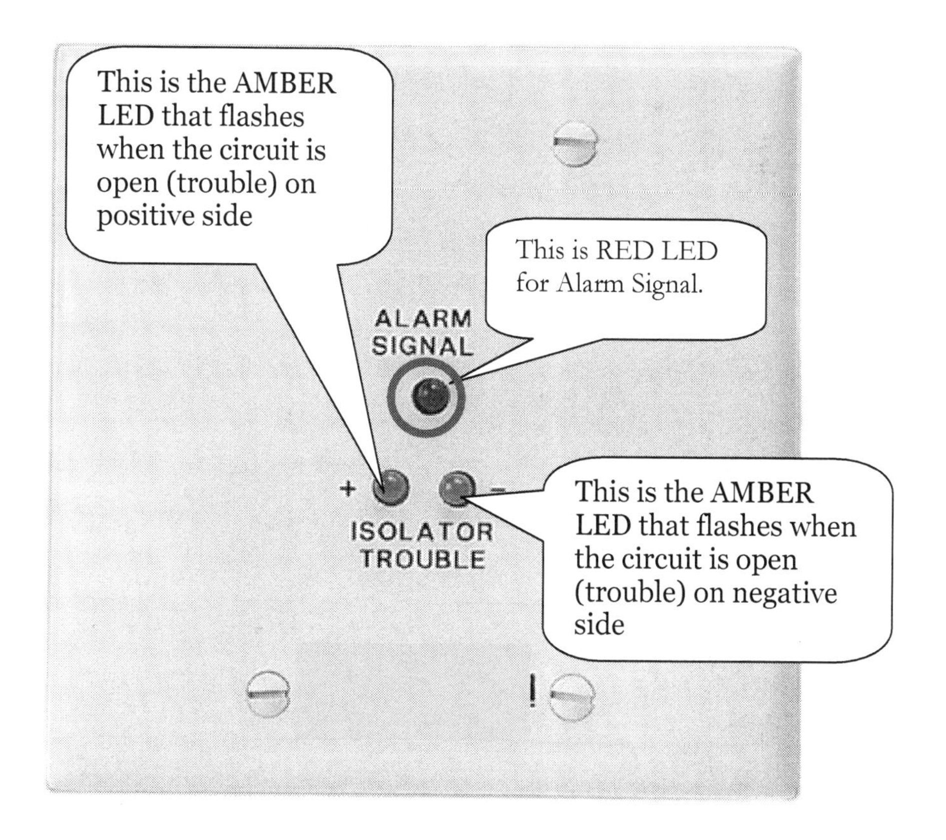

The device will fit into a 4" square electrical box 2.5" deep or larger. Anything that indicates the circuit's status should be installed in visible location that is also accessible for maintenance.Some advice for you as installer: make sure you test this device before installation and be attentive whilst choosing the size of the wiring since the terminal block will accommodate wiring no larger than #14 AWG(2.5 mmsq).Some designers require the silence switch installation to be done in the room where the horn is located. Most of the time, the horn is located 1'-00" under the ceiling line since the silence switch is 48 "high. I personally don't agree the installation of these silent switches since it allows people to ignore the fire alarm signal during a fire or emergency event.

The wiring diagram is shown in the next few illustrations.

There are different options:

- To install a SIGSM -100 that has a silence switch and a horn: The horn is located 1 foot under the ceiling line and the silence switch is accessible at 48 " above the finished floor (A.F.F.). This device requires 6 wires to be terminated.
- To install a silence-able Mini-Horn MH-S25: This mini-horn is provided with a silence button. This device requires 4 wires to be terminated.

## SCISM-Horn-Silence switch –Interconnections

Room 212 & Room 214 –Top view; with Horn & Silence Switch

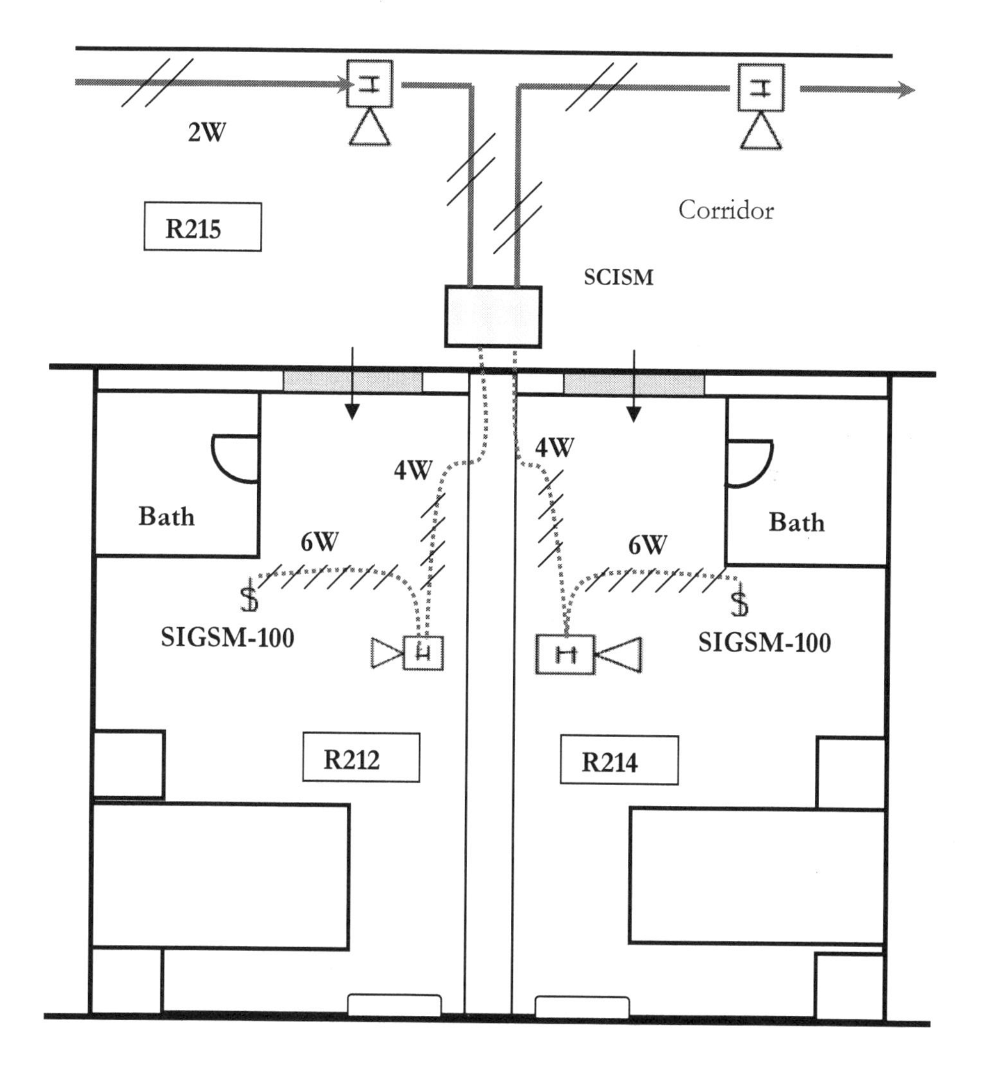


$ Silence switch **SCISM** Signal circuit isolator module

**SIGSM-100 (MIRCOM) will require 6 wires**

The silencing switch requires 6 wires to be terminated. It's better to check the terminal block of the device before installing the wiring. The wiring is protected in the EMT conduit by armored cable suitable for fire alarm system.

## SCISM-Horn –Interconnections

Room212 & Room 214; Top view, with silence-able Mini-Horn

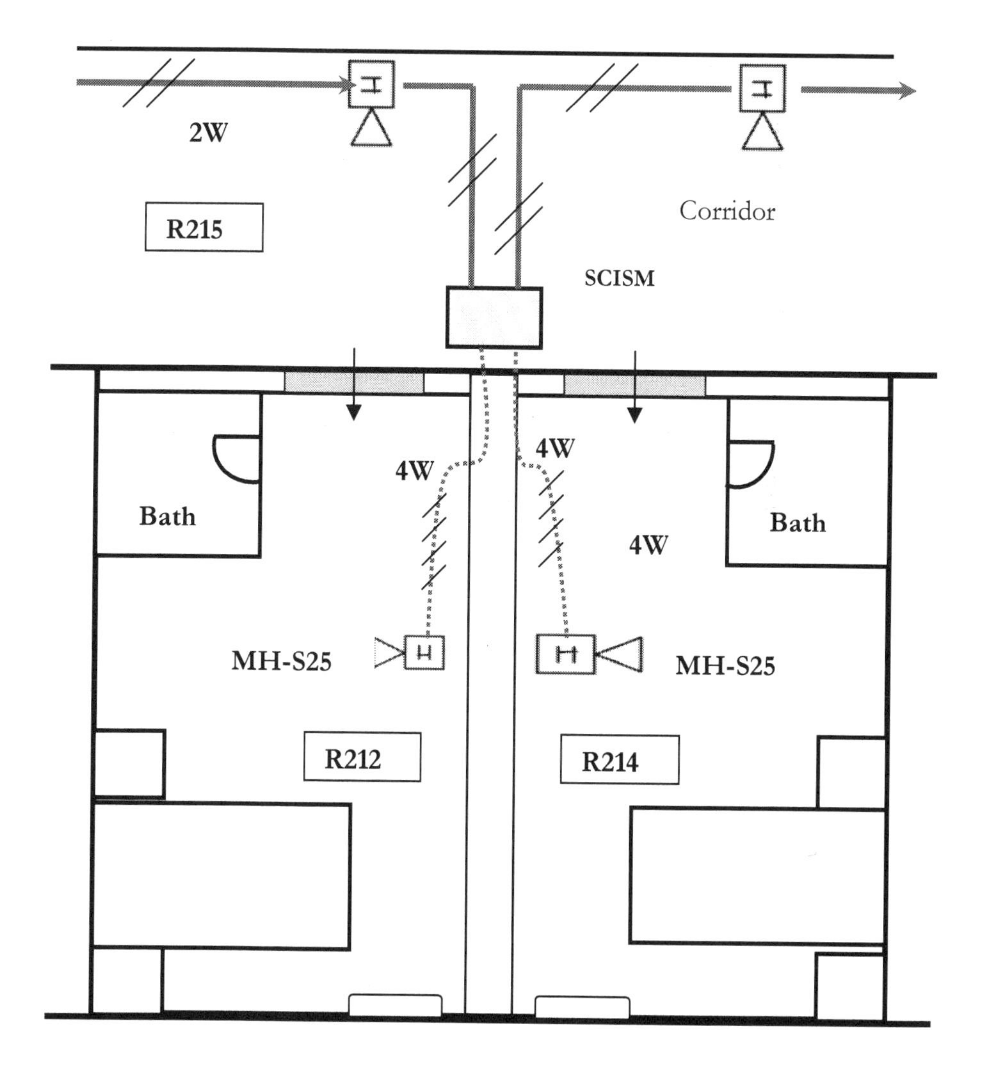


$ Silence switch; **SCISM** Signal circuit isolator module

The MH-S25 (MIRCOM) **Silence-able mini-horn** will require 4 wires. The Mini-Horn requires 4 wires to be terminated. It's best to check the terminal block of the device before installing the wiring. The wiring is protected into EMT conduit by armored cable suitable for the fire alarm system

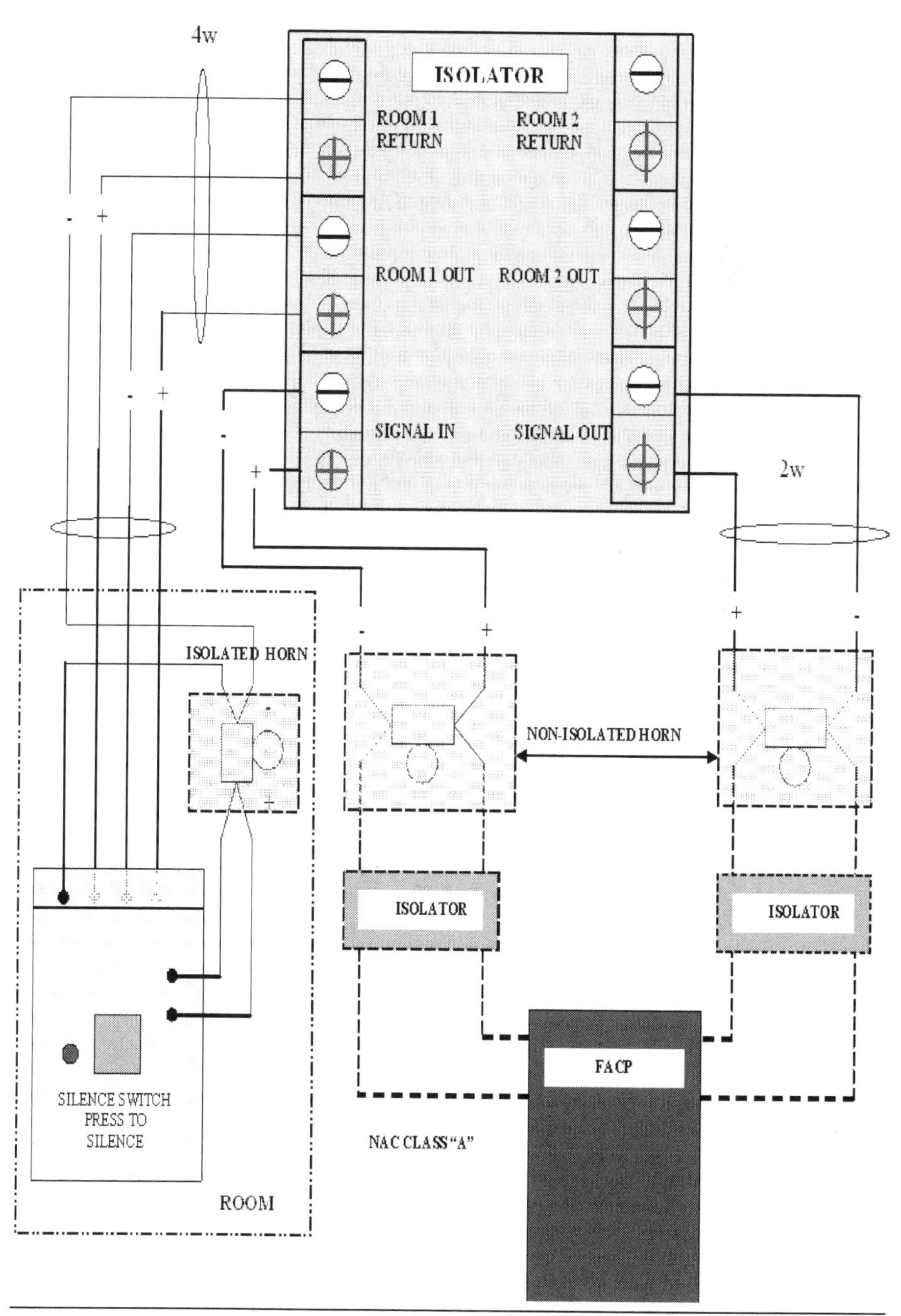
4w
ISOLATOR
ROOM 1 RETURN
ROOM 2 RETURN
ROOM 1 OUT
ROOM 2 OUT
SIGNAL IN
SIGNAL OUT
2w
ISOLATED HORN
NON-ISOLATED HORN
ISOLATOR
ISOLATOR
FACP
SILENCE SWITCH PRESS TO SILENCE
NAC CLASS "A"
ROOM

## The maximum number of devices on an NAC circuit

The devices on an NAC circuit (horn, combination horn & strobe, isolators, etc.) are "eating" more current than the devices we'd install on the INI (detectors, control modules, EOL resistors). The reason the devices from the NAC circuits are appreciable loads is because of the nature of loads. Basically they transform the electrical signal into both audible and visual signals. Compared with a detector or a control module, the amount of current is significantly lower. This means the number of devices should be carefully considered. In order to determine the maximum number of signaling appliances (NAC devices) that can be installed into a NAC circuit the total amount of current drawn by the signaling appliances should be calculated and adjusted in relation with the information from the manufacturer of the fire alarm control panel.

For example if you use a:

- "Fire Shield" 3 zones from Edwards, the maximum output power for NAC should not exceed 1.5 amps and the maximum power of the auxiliary devices (HVAC; Fire damper, etc.) should be max . 0.5 amps.
- "Fire Shield" 5 zones from Edwards- the maximum output power for NAC should not exceed 2.5 amps and the maximum power of auxiliary devices (HVAC; Fire damper, etc.) should be max. 0.5 amps.

- "Fire Shield" 10 zones from Edwards- the maximum output power for NAC should not exceed 5 amps and the maximum power of the auxiliary devices (HVAC; Fire damper, etc.) should be max. 0.5 amps.

For example, one horn 2447 TH (temporal horn) will draw about 20 mA while a 2452 THS (temporal horn and strobe) will draw 225 mA. So you can see the impact of such devices as a strobe. Using simple math, it is easy to establish the maximum number of a 2452 THS on NAC circuit when using a 3 zone Fire Shield fire alarm control panel:

1.5 Amps= 1500 mA. Now divide 1500 mA by 225 mA. This will result in about 6 ea

## More about FIRE ALARM SYSTEM

My next book to you is also about fire alarm system but this time we will go together through different learning system .I have a project of fire alarm system for your study so we will go through details this way to have our learning process more effective and efficient !

Next pages will display just a few pages to understand what all is about...............the new book!

UNDERSTANDING FIRE ALARM SYSTEM

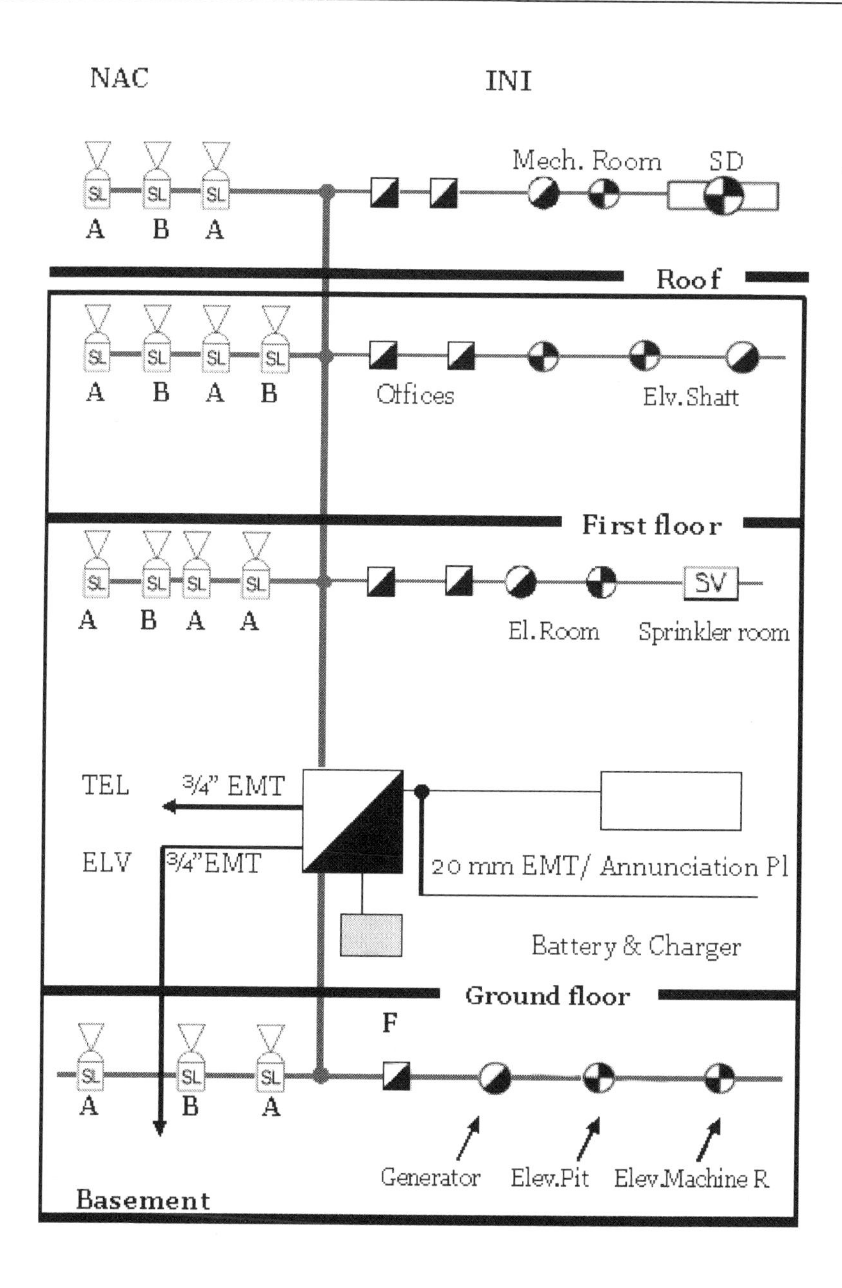


Block diagram for FAS and additional system

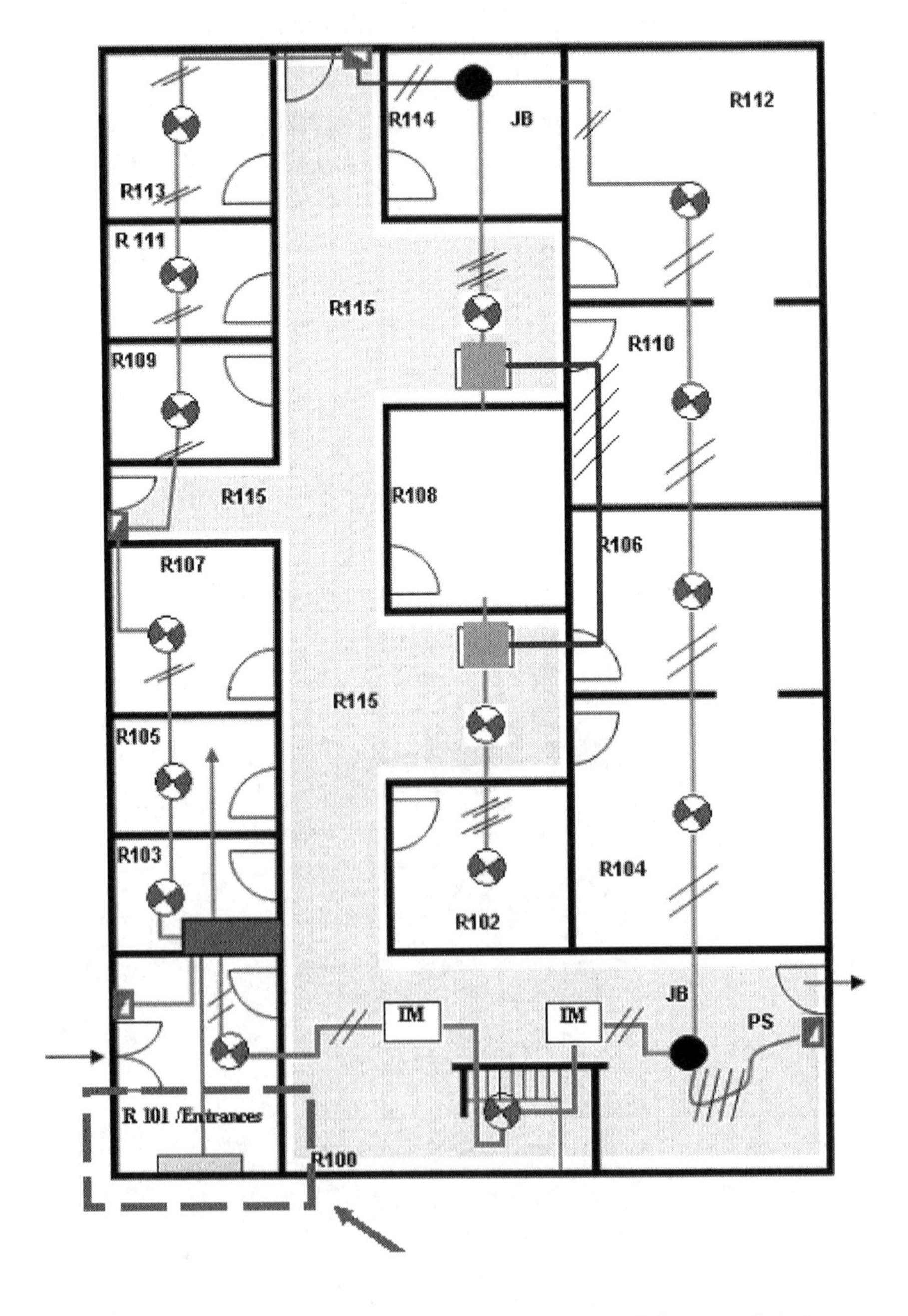


INITIATION CIRCUIT (detectors; pull stations; modules; conduits)

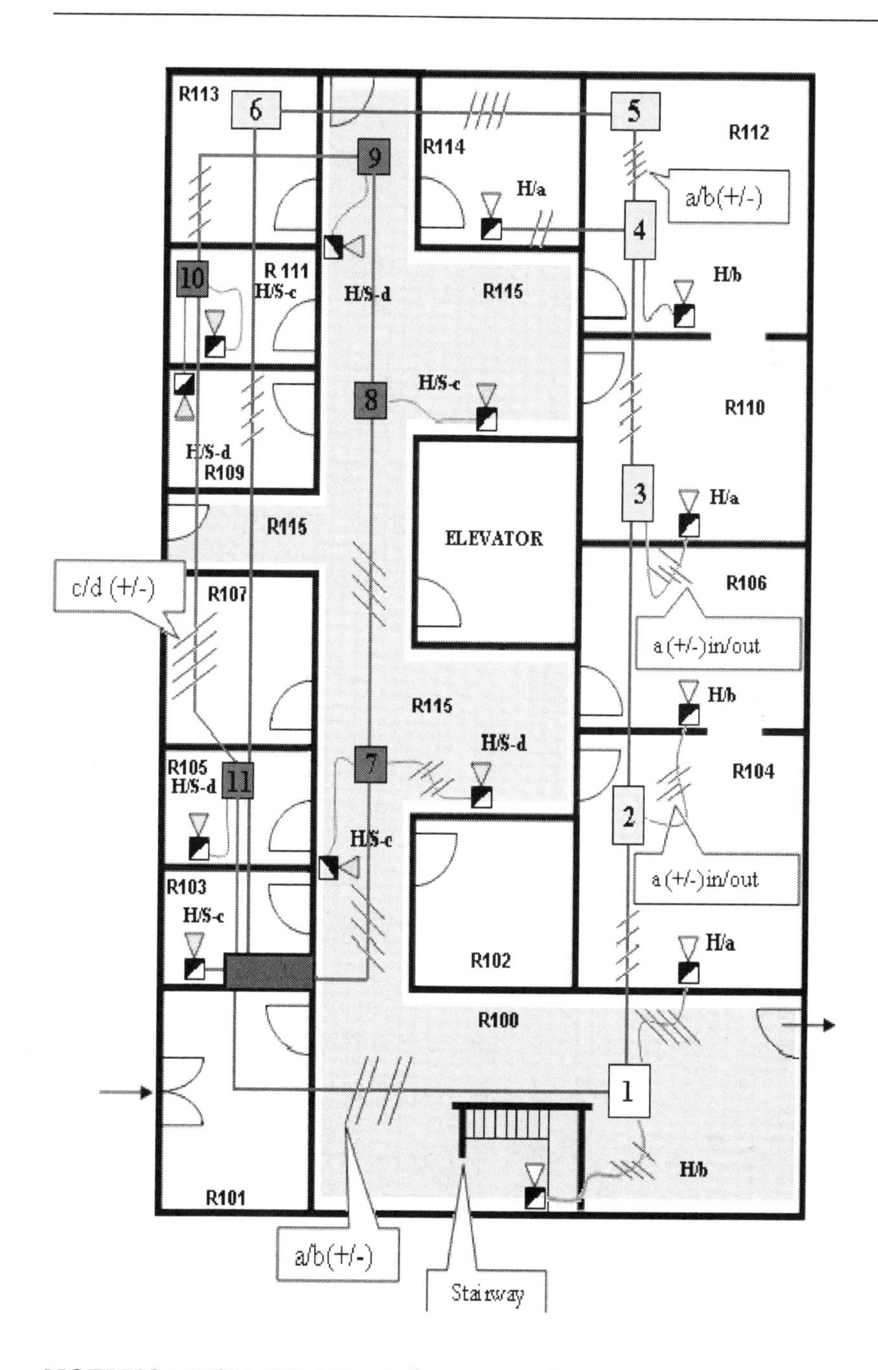


NOTIFICATION CIRCUITS (horns/strobes; conduits; boxes)

## Conduit system - typical detail installation

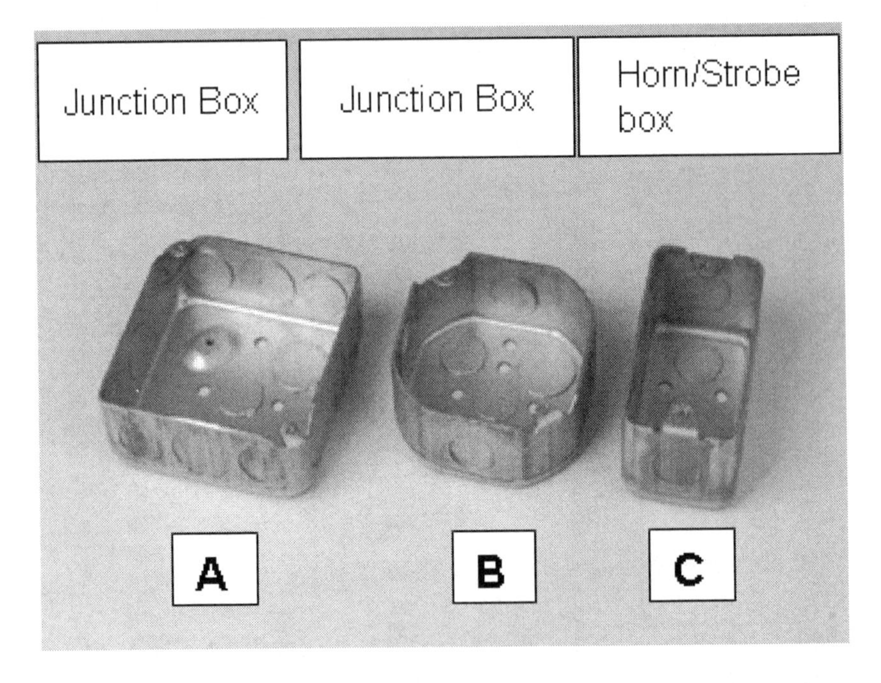


You've just been advised on how to deal with a situation when you should install, verify and/or design the correct type of boxes for devices on the initiation loop. Of course this is done in accordance with the environment. In our project we've examined the easiest case for a building like this (commercial building). You should expect to have different requirements if you need to install the fire alarm system inside a chemical plant, gas station or petrochemical plant. The system is identical but the designer will require and expect you to install heavy rigid conduit, special connectors, liquid tight metallic flexible and boxes or enclosures for the devices and panels that provide a high degree of protection. In other words, the level of hazard will dictate the installation details of the raceways. Let's go back to see the accessories:

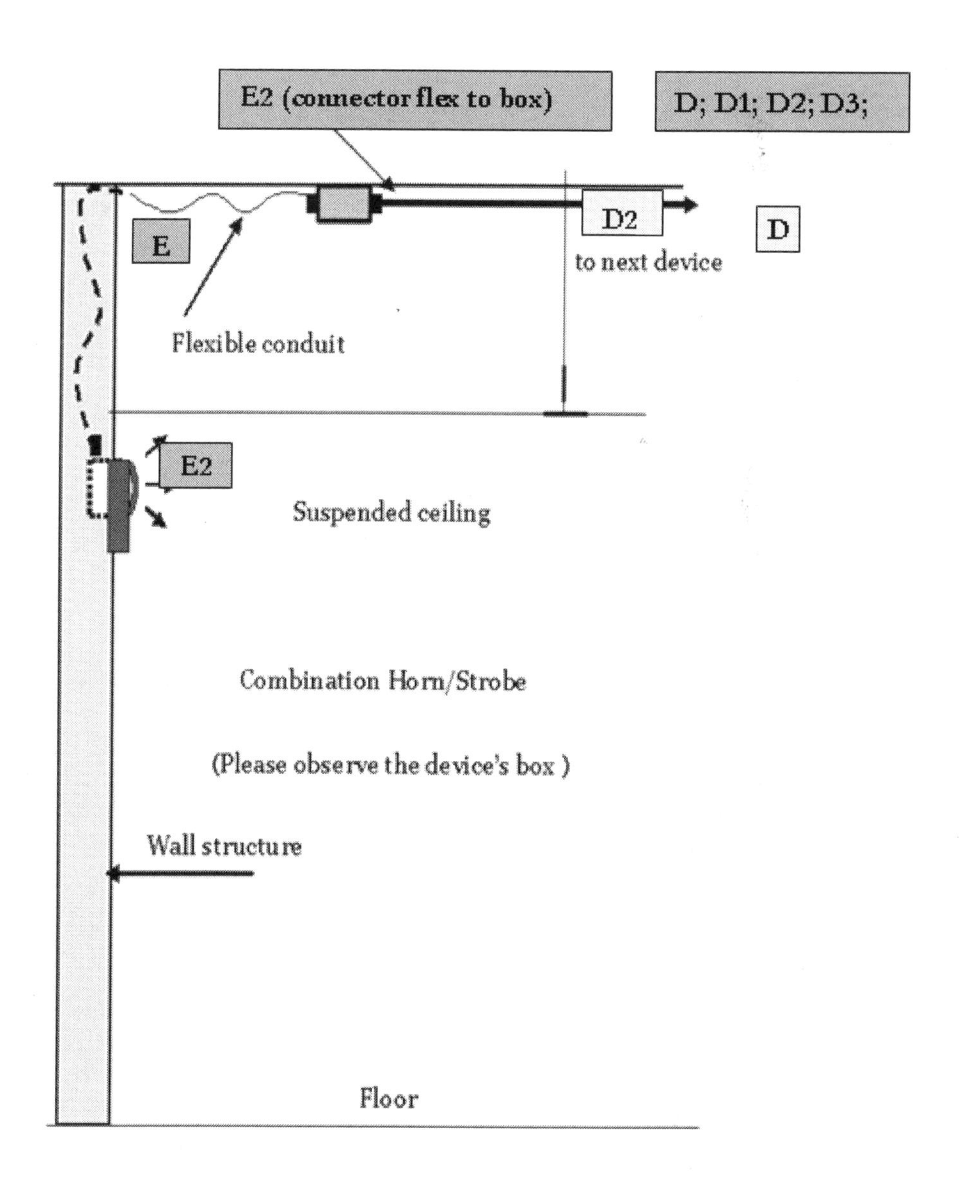

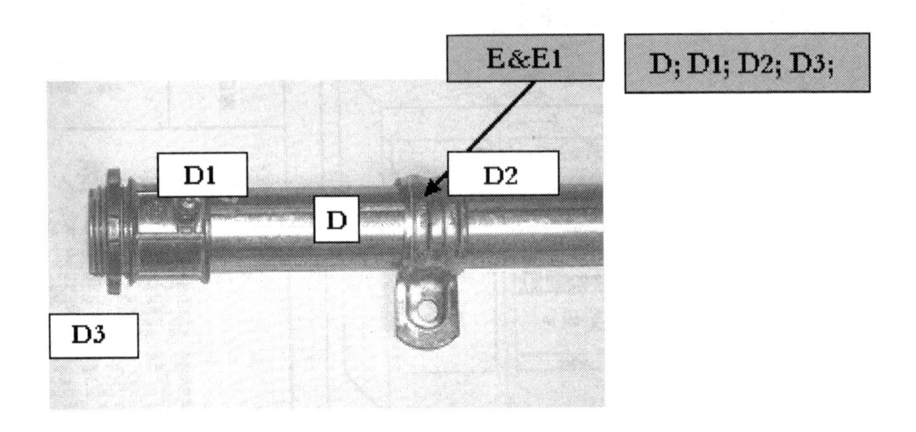


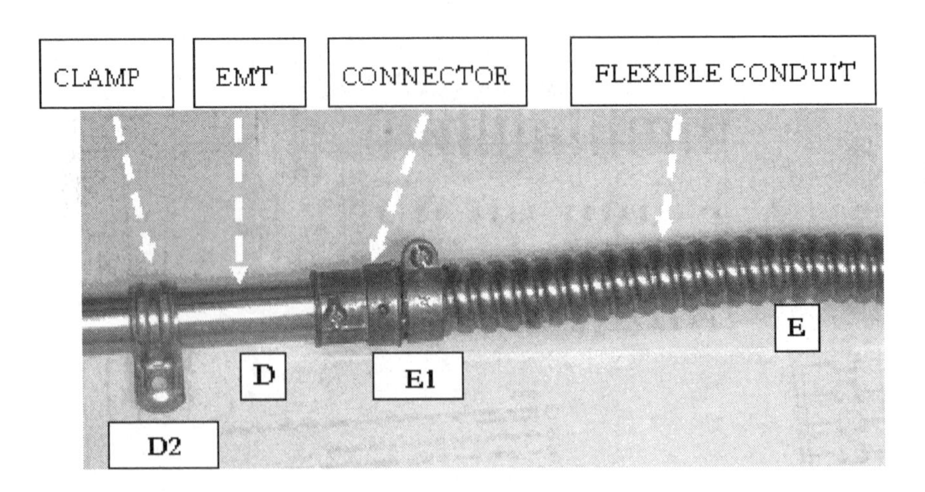


This shows flexible conduit connected to EMT conduit. Observe the right connector. The clamp is also visible.

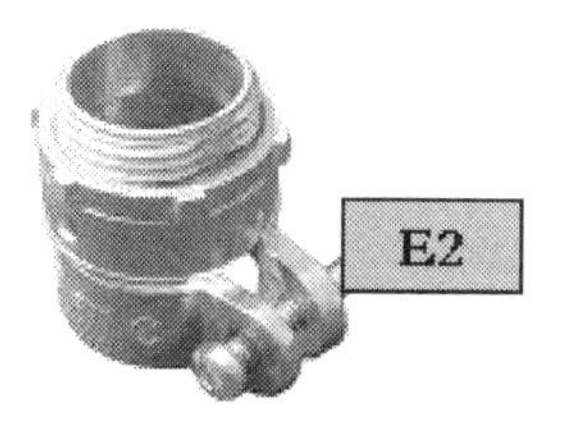


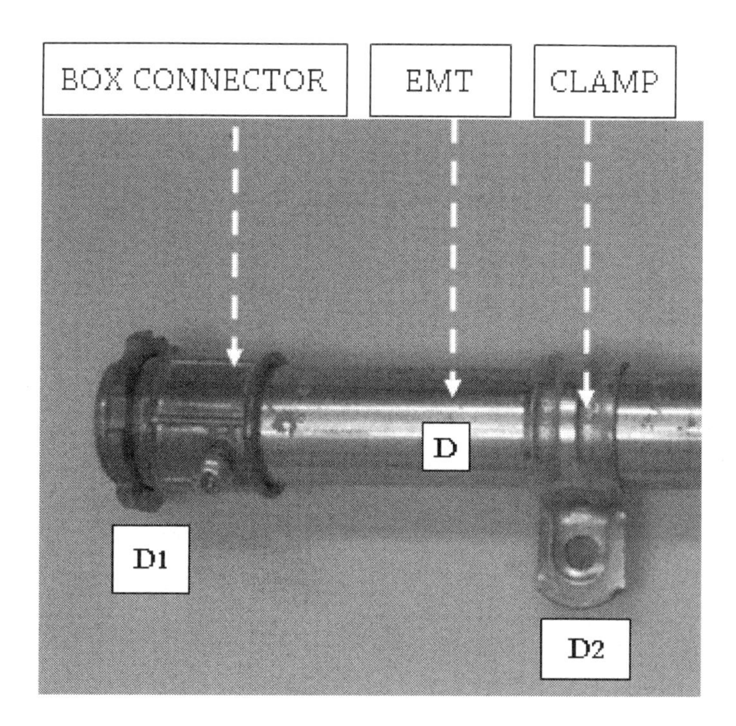


This type of connector will connect the EMT conduit to the flexible conduit or the flexible conduit to the junction box.

This connector E2 is connecting the rigid conduit to flexible conduit

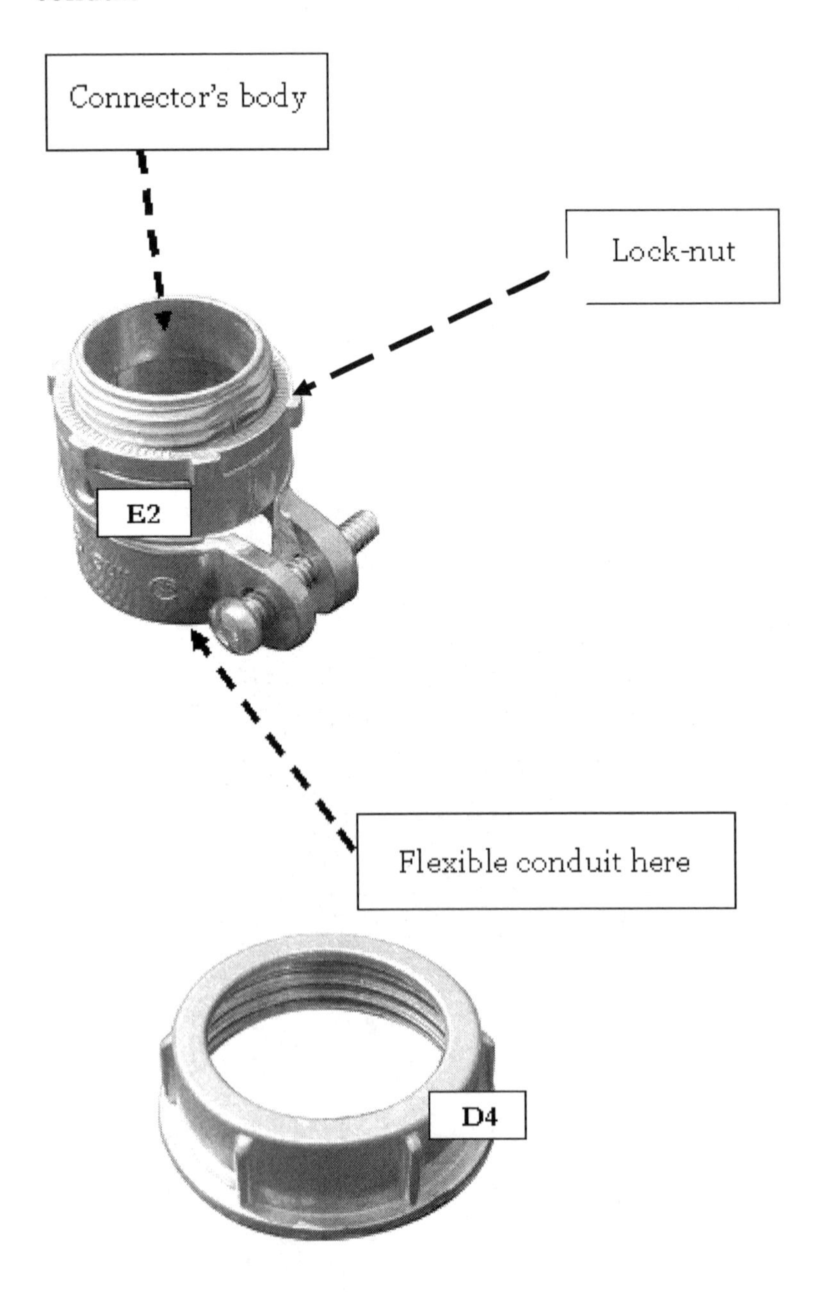

This is the "magic" PVC or Metal bushing. This is the item most commonly required by electrical inspectors in installations during the construction phase. It may look like an ordinary nut but its role protecting the wires in both the power circuit and signal circuit is a very important one. Having such bushing in place will ensure a path free of damages for wiring needing to be pulled into the EMT, conduit or RSG (rigid steel galvanized) conduit. This item needs to be installed on the connector or onto the conduit directly.

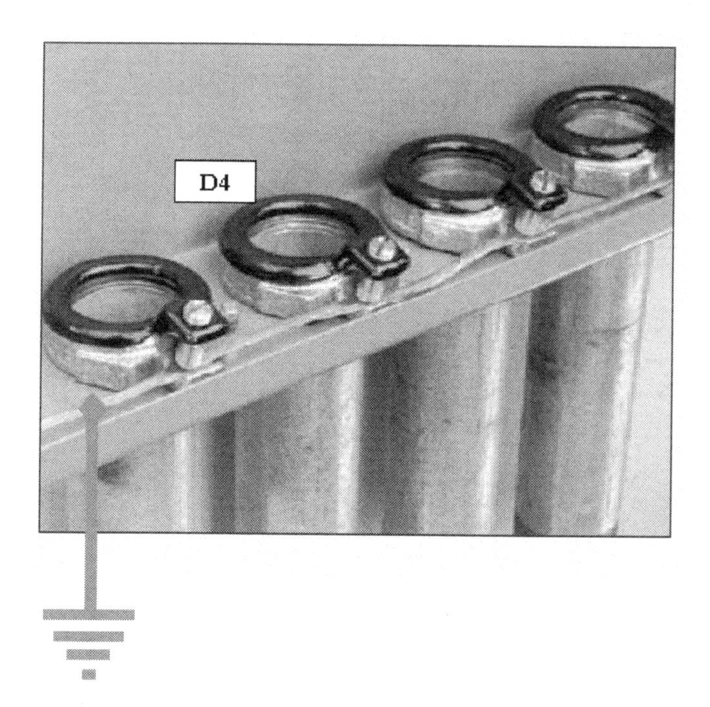


Observe the grounding clamp on the bushings and the bonding wire.

Combination Horn & Strobe, surface mounting details and conduit installation.

| ITEM | RACEWAY | SIZE | FORM | TO | ROOM |
|---|---|---|---|---|---|
| 1 | CONDUIT | 3/4" | FACP | BOX# 1 | R 100 |
| 2 | FLEXIBLE CONDUIT | 3/4" | BOX # 1 | H-b | Stairway (R 116) |
| 3 | FLEXIBLE CONDUIT | 3/4" | BOX # 1 | H-c | R 104 |
| 4 | CONDUIT | 3/4" | BOX # 1 | BOX # 2 | R 104 |
| 5 | FLEXIBLE CONDUIT | 3/4" | BOX # 2 | H-b | R 106 |
| 6 | CONDUIT | 3/4" | BOX # 2 | BOX # 3 | R 110 |
| 7 | FLEXIBLE CONDUIT | 3/4" | BOX # 3 | H-a | R 110 |
| 8 | CONDUIT | 3/4" | BOX # 3 | BOX # 4 | R110-R112 |
| 9 | FLEXIBLE CONDUIT | 3/4" | BOX # 4 | H-b | R 112 |
| 10 | FLEXIBLE CONDUIT | 3/4" | BOX # 4 | H-a | R 112-R114 |
| 11 | CONDUIT | 3/4" | BOX # 4 | BOX # 5 | R112 |
| 12 | CONDUIT | 3/4" | BOX # 5 | BOX # 6 | R112-R114-R115-R113 |
| 13 | CONDUIT | 3/4" | BOX # 6 | BOX #11 | R113-R111-R109-R115-R107-R105 |
| 14 | CONDUIT | 3/4" | BOX #11 | FACP | R 103 |

This table includes the items required to complete the installation of the raceway system for circuit "a "and circuit "b". If you attach a column with the quantity, this can be an MR (material request) or BOM (bill of materials). This is the prep work you need to do in order to ensure that the project is completed to the standards in quality and costs required by the designer.

***"Wiring shall be protected from mechanical injury or other injurious conditions such as moisture, excessive heat or corrosive action"***

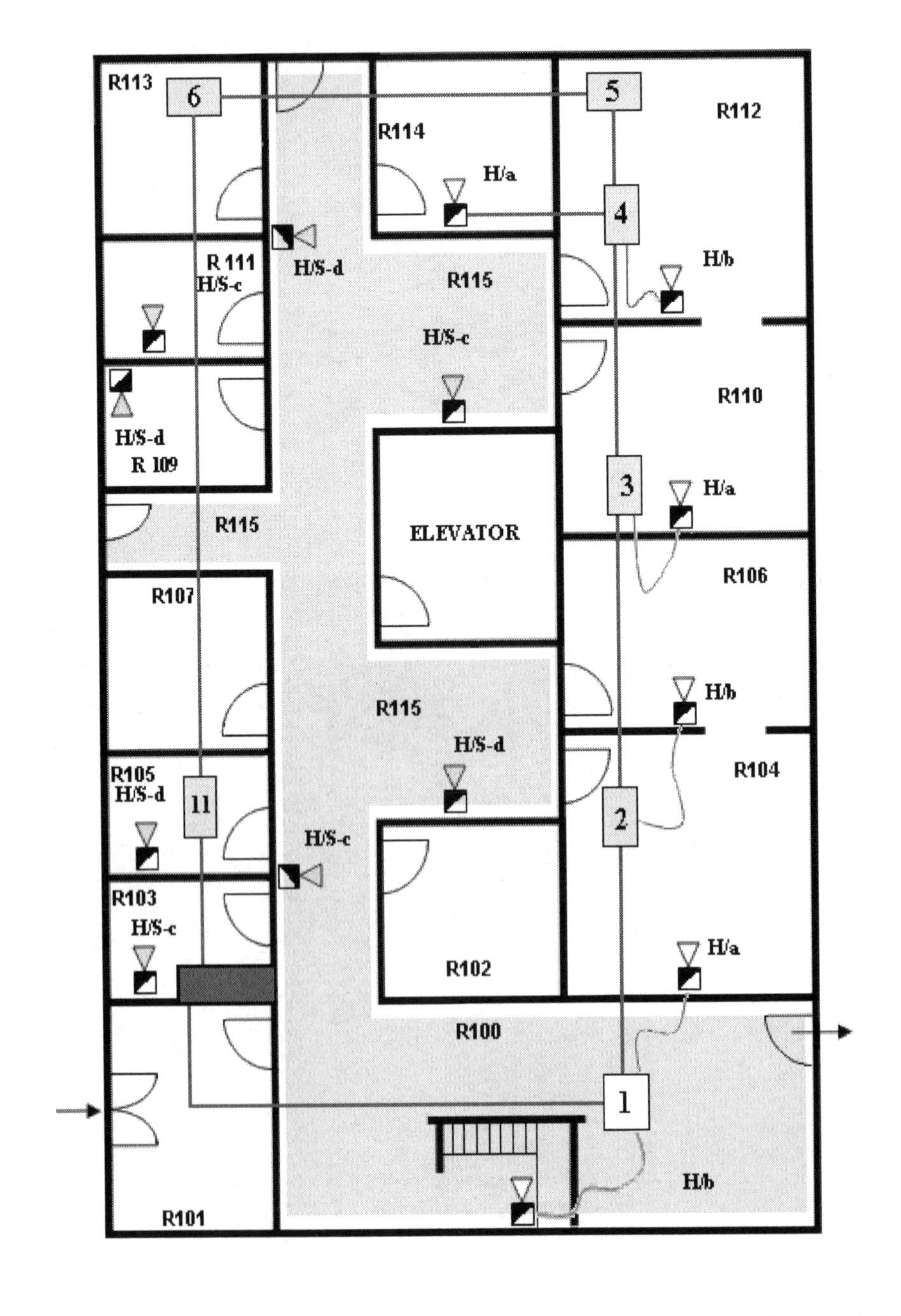
R113
6
5
R112
R114
H/a
4
R 111
H/S-c
H/S-d
R115
H/b
H/S-c
R110
H/S-d
R 109
3
H/a
R115
ELEVATOR
R106
R107
H/b
R115
H/S-d
R104
R105
H/S-d
11
2
H/S-c
R103
H/S-c
H/a
R102
R100
1
H/b
R101

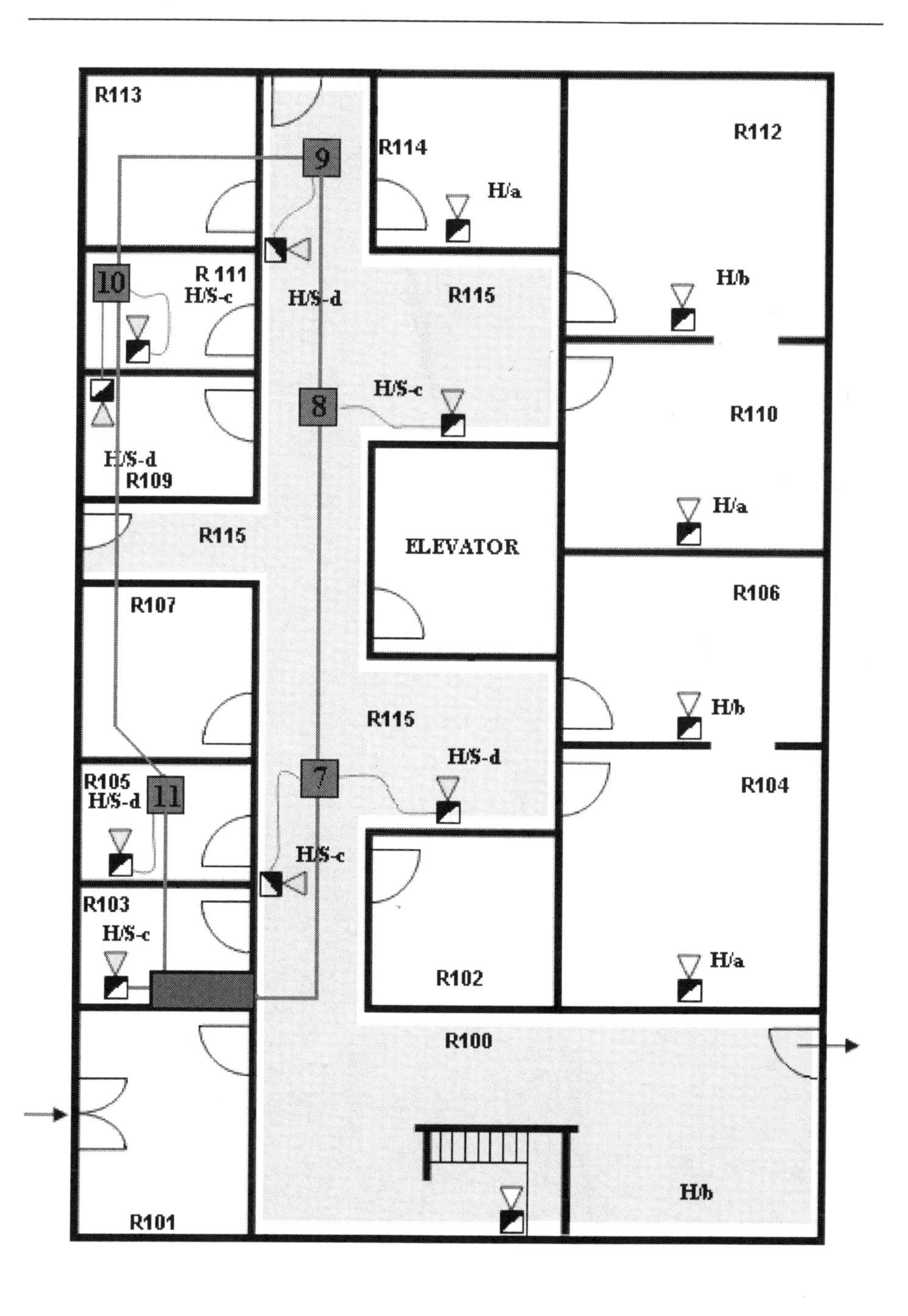
R113
9
R114
H/a
R112
10
R 111
H/S-c
H/S-d
R115
H/b
8
H/S-c
R110
H/S-d
R109
R115
ELEVATOR
H/a
R107
R106
H/b
R115
H/S-d
R104
R105
H/S-d
7
11
H/S-c
R103
H/S-c
R102
H/a
R100
H/b
R101

Made in the USA
Middletown, DE
10 September 2016